Advanced Level Mathematics
Statistics 2

Steve Dobbs and Jane Miller

CAMBRIDGE
UNIVERSITY PRESS

CAMBRIDGE
UNIVERSITY PRESS

University Printing House, Cambridge CB2 8BS, United Kingdom

Cambridge University Press is part of the University of Cambridge.

It furthers the University's mission by disseminating knowledge in the pursuit of education, learning and research at the highest international levels of excellence.

www.cambridge.org
Information on this title: www.cambridge.org/9780521530149

© Cambridge University Press 2003

First published 2003
13th printing 2014

Printed in Italy by Rotolito Lombarda S.p.A.

A catalogue record for this publication is available from the British Library

ISBN 978-0-521-53014-9 Paperback

ACKNOWLEDGEMENTS

The publishers would like to acknowledge the contributions of the following people to this series of books: Tim Cross, Richard Davies, Maurice Godfrey, Chris Hockley, Lawrence Jarrett, David A. Lee, Jean Matthews, Norman Morris, Charles Parker, Geoff Staley, Rex Stephens, Peter Thomas and Owen Toller.

Cover image: © Tony Stone Images / Mark Harwood

Contents

Introduction

University of Cambridge International Examinations (CIE) Advanced Level Mathematics has been written especially for the new CIE mathematics syllabus. There is one book corresponding to each syllabus unit, except that units P2 and P3 are contained in a single book. This book is the second Probability and Statistics unit, S2.

The syllabus content is arranged by chapters which are ordered so as to provide a viable teaching course. A few sections include important results that are difficult to prove or outside the syllabus. These sections are marked with an asterisk (*) in the section heading, and there is usually a sentence early on explaining precisely what it is that the student needs to know.

Some paragraphs within the text appear in *this type style*. These paragraphs are usually outside the main stream of the mathematical argument, but may help to give insight, or suggest extra work or different approaches.

Graphic calculators are not permitted in the examination, but they can be useful aids in learning mathematics. In the book the authors have noted where access to graphic calculators would be especially helpful but have not assumed that they are available to all students.

The authors have assumed that students have access to calculators with built-in statistical functions.

Numerical work is presented in a form intended to discourage premature approximation. In ongoing calculations inexact numbers appear in decimal form like $3.456\ldots$, signifying that the number is held in a calculator to more places than are given. Numbers are not rounded at this stage; the full display could be either $3.456\,123$ or $3.456\,789$. Final answers are then stated with some indication that they are approximate, for example '1.23 correct to 3 significant figures'.

Most chapters contain Practical activities. These can be used either as an introduction to a topic, or, later on, to reinforce the theory. Two Practical activities, in Sections 4.5 and 5.4, require access to a computer.

There are also plenty of exercises, and each chapter ends with a Miscellaneous exercise which includes some questions of examination standard. There is a Revision exercise, and two Practice examination papers. In some exercises a few of the later questions may go beyond the likely requirements of the examination, either in difficulty or in length or both. Some questions are marked with an asterisk, which indicates that they require knowledge of results outside the syllabus.

Cambridge University Press would like to thank OCR (Oxford, Cambridge and RSA Examinations), part of the University of Cambridge Local Examinations Syndicate (UCLES) group, for permission to use past examination questions set in the United Kingdom.

The authors thank CIE and Cambridge University Press for their help in producing this book. However, the responsibility for the text, and for any errors, remains with the authors.

1 The Poisson distribution

This chapter introduces a discrete probability distribution which is used for modelling random events. When you have completed it you should

- be able to calculate probabilities for the Poisson distribution
- understand the relevance of the Poisson distribution to the distribution of random events and use the Poisson distribution as a model
- be able to use the result that the mean and variance of a Poisson distribution are equal
- be able to use the Poisson distribution as an approximation to the binomial distribution where appropriate
- be able to use the normal distribution, with a continuity correction, as an approximation to the Poisson distribution where appropriate.

1.1 The Poisson probability formula

Situations often arise where the variable of interest is the number of occurrences of a particular event in a given interval of space or time. An example is given in Table 1.1. This shows the frequency of 0, 1, 2 etc. phone calls arriving at a switchboard in 100 consecutive time intervals of 5 minutes. In this case the 'event' is the arrival of a phone call and the 'given interval' is a time interval of 5 minutes.

Number of calls	0	1	2	3	4 or more
Frequency	71	23	4	2	0

Table 1.1. Frequency distribution of number of telephone calls in 5-minute intervals.

Some other examples are

- the number of cars passing a point on a road in a time interval of 1 minute,
- the number of misprints on each page of a book,
- the number of radioactive particles emitted by a radioactive source in a time interval of 1 second.

Further examples can be found in the practical activities in Section 1.4.

The probability distribution which is used to model these situations is called the **Poisson distribution** after the French mathematician and physicist Siméon-Denis Poisson (1781–1840). The distribution is defined by the probability formula

$$P(X = x) = e^{-\lambda} \frac{\lambda^x}{x!}, \qquad x = 0, 1, 2, \ldots .$$

This formula involves the mathematical constant e *which you may have already met in unit P2. If you have not, then it is enough for you to know at this stage that the approximate value of* e *is* 2.718 *and that powers of* e *can be found using your calculator.*

Check that you can use your calculator to show that $e^{-2} = 0.135...$ *and* $e^{-0.1} = 0.904....$

The method by which Poisson arrived at this formula will be outlined in Section 1.2.

This formula involves only one parameter, λ. (λ, pronounced 'lambda', is the Greek letter l.) You will see later that λ is the mean of the distribution. The notation for indicating that a random variable X has a Poisson distribution with mean λ is $X \sim \text{Po}(\lambda)$. Once λ is known you can calculate $P(X = 0)$, $P(X = 1)$ etc. There is no upper limit on the value of X.

Example 1.1.1

The number of particles emitted per second by a radioactive source has a Poisson distribution with mean 5. Calculate the probabilities of
(a) 0,   (b) 1,   (c) 2,   (d) 3 or more emissions in a time interval of 1 second.

    (a) Let X be the random variable 'the number of particles emitted in 1 second'. Then

$$X \sim \text{Po}(5). \text{ Using the Poisson probability formula } P(X = x) = e^{-\lambda}\frac{\lambda^x}{x!} \text{ with } \lambda = 5,$$

$$P(X = 0) = e^{-5}\frac{5^0}{0!} = 0.006\ 737... = 0.006\ 74, \text{ correct to 3 significant figures.}$$

Recall that $0! = 1$ *(see P1 Section 8.3).*

    (b) $P(X = 1) = e^{-5}\dfrac{5^1}{1!} = 0.033\ 68... = 0.0337$, correct to 3 significant figures.

    (c) $P(X = 2) = e^{-5}\dfrac{5^2}{2!} = 0.084\ 22... = 0.0842$, correct to 3 significant figures.

    (d) Since there is no upper limit on the value of X the probability of 3 or more emissions must be found by subtraction.

$$P(X \geqslant 3) = 1 - P(X = 0) - P(X = 1) - P(X = 2)$$
$$= 1 - 0.006\ 737... - 0.033\ 68... - 0.084\ 22...$$
$$= 0.875, \text{correct to 3 significant figures.}$$

Example 1.1.2

The number of demands for taxis to a taxi firm is Poisson distributed with, on average, four demands every 30 minutes. Find the probabilities of
(a) no demand in 30 minutes,
(b) 1 demand in 1 hour,
(c) fewer than 2 demands in 15 minutes.

    (a) Let X be the random variable 'the number of demands in a 30 minute interval'. Then $X \sim \text{Po}(4)$. Using the Poisson formula with $\lambda = 4$,

$$P(X = 0) = e^{-4}\frac{4^0}{0!} = 0.0183, \text{ correct to 3 significant figures.}$$

(b) Let Y be the random variable 'the number of demands in a 1 hour interval'. As the time interval being considered has changed from 30 minutes to 1 hour, you must change the value of λ to equal the mean for this new time interval, that is to 8, giving $Y \sim \text{Po}(8)$. Using the Poisson formula with $\lambda = 8$,

$$P(Y = 1) = e^{-8} \frac{8^1}{1!} = 0.002\,68 \text{, correct to 3 significant figures.}$$

(c) Again the time interval has been altered. Now the appropriate value for λ is 2. Let W be the number of demands in 15 minutes. Then $W \sim \text{Po}(2)$.

$$P(W < 2) = P(W = 0) + P(W = 1) = e^{-2} \frac{2^0}{0!} + e^{-2} \frac{2^1}{1!} = 0.406,$$

correct to 3 significant figures.

Here is a summary of the results of this section.

The **Poisson distribution** is used as a model for the number, X, of events in a given interval of space or time. It has the probability formula

$$P(X = x) = e^{-\lambda} \frac{\lambda^x}{x!}, \qquad x = 0, 1, 2, \ldots,$$

where λ is equal to the mean number of events in the given interval.

The notation $X \sim \text{Po}(\lambda)$ indicates that X has a Poisson distribution with mean λ.

Some books use μ rather than λ to denote the parameter of a Poisson distribution.

Exercise 1A

1   The random variable T has a Poisson distribution with mean 3. Calculate

   (a) $P(T = 2)$,   (b) $P(T \leqslant 1)$,   (c) $P(T \geqslant 3)$.

2   Given that $U \sim \text{Po}(3.25)$, calculate

   (a) $P(U = 3)$,   (b) $P(U \leqslant 2)$,   (c) $P(U \geqslant 2)$.

3   The random variable W has a Poisson distribution with mean 2.4. Calculate

   (a) $P(W \leqslant 3)$,   (b) $P(W \geqslant 2)$,   (c) $P(W = 3)$.

4   Accidents on a busy urban road occur at a mean rate of 2 per week. Assuming that the number of accidents per week follows a Poisson distribution, calculate the probability that

   (a) there will be no accidents in a particular week,

   (b) there will be exactly 2 accidents in a particular week,

   (c) there will be fewer than 3 accidents in a given two-week period.

5 On average, 15 customers a minute arrive at the check-outs of a busy supermarket. Assuming that a Poisson distribution is appropriate, calculate

(a) the probability that no customers arrive at the check-outs in a given 10-second interval,

(b) the probability that more than 3 customers arrive at the check-outs in a 15-second interval,

6 During April of this year, Malik received 15 telephone calls. Assuming that the number of telephone calls he receives in April of next year follows a Poisson distribution with the same mean number of calls per day, calculate the probability that

(a) on a given day in April next year he will receive no telephone calls,

(b) in a given 7-day week next April he will receive more than 3 telephone calls.

7 Assume that cars pass under a bridge at a rate of 100 per hour and that a Poisson distribution is appropriate.

(a) What is the probability that during a 3-minute period no cars will pass under the bridge?

(b) What time interval is such that the probability is at least 0.25 that no car will pass under the bridge during that interval?

8 A radioactive source emits particles at an average rate of 1 per second. Assume that the number of emissions follows a Poisson distribution.

(a) Calculate the probability that 0 or 1 particle will be emitted in 4 seconds.

*(b) The emission rate changes such that the probability of 0 or 1 emission in 4 seconds becomes 0.8. What is the new emission rate?

1.2 Modelling random events

The examples which you have already met in this chapter have assumed that the variable you are dealing with has a Poisson distribution. How can you decide whether the Poisson distribution is a suitable model if you are not told? The answer to this question can be found by considering the way in which the Poisson distribution is related to the binomial distribution in the situation where the number of trials is very large and the probability of success is very small.

Table 1.2 reproduces Table 1.1 giving the frequency distribution of phone calls in 100 5-minute intervals.

Number of calls	0	1	2	3	4 or more
Frequency	71	23	4	2	0

Table 1.2. Frequency distribution of number of telephone calls in 5-minute intervals.

If these calls were plotted on a time axis you might see something which looked like Fig. 1.3.

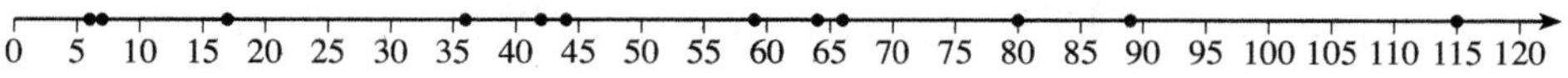

Fig. 1.3. Times of arrival of telephone calls at a switchboard.

The time axis has been divided into 5-minute intervals (only 24 are shown) and these intervals can contain $0, 1, 2$ etc. phone calls. Suppose now that you assume that the phone calls occur *independently* of each other and *randomly* in time. In order to make the terms in italics clearer consider the following. Imagine the time axis is divided up into very small intervals of width δt (where δ is used in the same way as it is in pure mathematics). These intervals are so small that they never contain more than one call. If the calls are *random* then the probability that one of these intervals contains a call does not depend on which interval is considered; that is, it is constant. If the calls are *independent* then whether or not one interval contains a call has no effect on whether any other interval contains a call.

Looking at each interval of width δt in turn to see whether it contains a call or not gives a series of trials, each with two possible outcomes. This is just the kind of situation which is described by the binomial distribution (see S1 Chapter 7). These trials also satisfy the conditions for the binomial distribution that they should be independent and have a fixed probability of success.

Suppose that a 5-minute interval contains n intervals of width δt. If there are, on average, λ calls every 5 minutes then the proportion of intervals which contain a call will be equal to $\dfrac{\lambda}{n}$. The probability, p, that one of these intervals contains a call is therefore equal to $\dfrac{\lambda}{n}$. Since δt is small, n is large and $\dfrac{\lambda}{n}$ is small. You can verify from Table 1.2 that the mean number of calls in a 5-minute interval is 0.37 so the distribution of X, the number of calls in a 5-minute interval, is $B\left(n, \dfrac{0.37}{n}\right)$.

Finding $P(X = 0)$ Using the binomial probability formula $P(X = x) = \dbinom{n}{x} p^x q^{n-x}$,

you can calculate, for example, the probability of zero calls in a 5-minute interval as

$$P(X = 0) = \binom{n}{0}\left(\frac{0.37}{n}\right)^0 \left(1 - \frac{0.37}{n}\right)^n.$$

In order to proceed you need a value for n. Recall that n must be large enough to ensure that the δt-intervals never contain more than one call. Suppose $n = 1000$. This gives

$$P(X = 0) = \binom{1000}{0}\left(\frac{0.37}{1000}\right)^0 \left(1 - \frac{0.37}{1000}\right)^{1000} = 0.690\,68\dots.$$

However, even with such a large number of intervals there is still a chance that one of the δt-intervals could contain more than one call, so a larger value of n would be better. Try $n = 10\,000$ giving

$$P(X = 0) = \binom{10\,000}{0}\left(\frac{0.37}{10\,000}\right)^0 \left(1 - \frac{0.37}{10\,000}\right)^{10000} = 0.690\,72\dots.$$

Explore for yourself what happens as you increase the value of n still further. You should find that your answers tend towards the value $0.690\,73\ldots$. This is equal to $e^{-0.37}$, which is the value the Poisson probability formula gives for $P(X=0)$ when $\lambda = 0.37$.

This is an example of the general result that $\left(1-\dfrac{x}{n}\right)^n$ tends to the value e^{-x} as n tends to infinity.

Provided that two events cannot occur simultaneously, allowing n to tend to infinity will ensure that not more than one event can occur in a δt-interval.

Finding $P(X=1)$ In a similar way you can find the probability of one call in a 5-minute interval by starting from the binomial formula and allowing n to increase as follows.

$$P(X=1) = \binom{n}{1}\left(\frac{0.37}{n}\right)^1\left(1-\frac{0.37}{n}\right)^{n-1} = 0.37\left(1-\frac{0.37}{n}\right)^{n-1}.$$

Putting $n=1000$,

$$P(X=1) = 0.37\left(1-\frac{0.37}{1000}\right)^{999} = 0.37\times 0.690\,94\ldots = 0.255\,64\ldots.$$

Putting $n=10\,000$,

$$P(X=1) = 0.37\left(1-\frac{0.37}{10\,000}\right)^{9999} = 0.37\times 0.690\,75\ldots = 0.255\,579\ldots.$$

Again, you should find that, as n increases, the probability tends towards the value given by the Poisson probability formula,

$$P(X=1) = 0.37\times e^{-0.37} = 0.255\,57\ldots.$$

Finding $P(X=2)$, $P(X=3)$, etc. You could verify for yourself that similar results are obtained when the probabilities of $X=2,3$, etc. are calculated by a similar method. A spreadsheet program or a programmable calculator would be helpful.

The general result for $P(X=x)$ can be derived as follows. Starting with $X \sim B\left(n,\dfrac{\lambda}{n}\right)$.

$$P(X=x) = \binom{n}{x}\left(\frac{\lambda}{n}\right)^x\left(1-\frac{\lambda}{n}\right)^{n-x} = \frac{n(n-1)(n-2)\ldots(n-x+1)}{x!}\times\frac{\lambda^x}{n^x}\left(1-\frac{\lambda}{n}\right)^{n-x}$$

$$= \frac{\lambda^x}{x!}\times\frac{n-1}{n}\times\frac{n-2}{n}\times\ldots\times\frac{n-x+1}{n}\times\left(1-\frac{\lambda}{n}\right)^{n-x}.$$

Now consider what happens as n gets larger. The fractions $\dfrac{n-1}{n}, \dfrac{n-2}{n}$, etc. tend towards 1. The term $\left(1-\dfrac{\lambda}{n}\right)^{n-x}$ can be approximated by $\left(1-\dfrac{\lambda}{n}\right)^{n}$ since x, a constant, is negligible compared with n and, as you have seen previously, this tends towards $e^{-\lambda}$.

Combining these results gives

$$P(X = x) = \frac{\lambda^x}{x!} e^{-\lambda}.$$

The assumptions made in the derivation above give the conditions that a set of events must satisfy for the Poisson distribution to be a suitable model. They are listed below.

> The Poisson distribution is a suitable model for events which
>
> - occur randomly in space or time,
> - occur singly, that is events cannot occur simultaneously,
> - occur independently, and
> - occur at a constant rate, that is the mean number of events in a given time interval is proportional to the size of the interval.

Example 1.2.1

For each of the following situations state whether the Poisson distribution would provide a suitable model. Give reasons for your answers.

(a) The number of cars per minute passing under a road bridge between 10 a.m. and 11 a.m. when the traffic is flowing freely.

(b) The number of cars per minute entering a city-centre carpark on a busy Saturday between 9 a.m. and 10 a.m.

(c) The number of particles emitted per second by a radioactive source.

(d) The number of currants in buns sold at a particular baker's shop on a particular day.

(e) The number of blood cells per ml in a dilute solution of blood which has been left standing for 24 hours.

(f) The number of blood cells per ml in a well-shaken dilute solution of blood.

(a) The Poisson distribution should be a good model for this situation as the appropriate conditions should be met: since the traffic is flowing freely the cars should pass independently and at random; it is not possible for cars to pass simultaneously; the average rate of traffic flow is likely to be constant over the time interval given.

(b) The Poisson distribution is unlikely to be a good model: if it is a busy day the cars will be queuing for the carpark and so they will not be moving independently.

(c) The Poisson distribution should be a good model provided that the time period over which the measurements are made is much longer than the lifetime of the source: this will ensure that the average rate at which the particles are emitted is constant. Radioactive particles are emitted independently and at random and, for practical purposes, they can be considered to be emitted singly.

(d) The Poisson distribution should be a good model provided that the following conditions are met: all the buns are prepared from the same mixture so that the average number of currants per bun is constant; the mixture is well stirred so that the currants are distributed at random; the currants do not stick to each other or touch each other so that they are positioned independently.

(e) The Poisson distribution will not be a good model because the blood cells will have tended to sink towards the bottom of the solution. Thus the average number of blood cells per ml will be greater at the bottom than the top.

(f) If the solution has been well shaken the Poisson distribution will be a suitable model. The blood cells will be distributed at random and at a constant average rate. Since the solution is dilute the blood cells will not be touching and so will be positioned independently.

1.3   The variance of a Poisson distribution

In Section 1.2 the Poisson probability formula was deduced from the distribution of $X \sim B\left(n, \dfrac{\lambda}{n}\right)$ by considering what happens as n tends to infinity. The variance of a Poisson distribution can be obtained by considering what happens to the variance of the distribution of $X \sim B\left(n, \dfrac{\lambda}{n}\right)$ as n gets very large. In S1 Section 8.3 you met the formula $\text{Var}(X) = npq$ for the variance of a binomial distribution. Substituting for p and q gives

$$\text{Var}(X) = n \times \frac{\lambda}{n}\left(1 - \frac{\lambda}{n}\right) = \lambda\left(1 - \frac{\lambda}{n}\right).$$

As n gets very large the term $\dfrac{\lambda}{n}$ tends to zero. This gives λ as the variance of the Poisson distribution. Thus the Poisson distribution has the interesting property that its mean and variance are equal.

> For a Poisson distribution $X \sim \text{Po}(\lambda)$
>
> $\qquad$ mean $= \mu = \text{E}(X) = \lambda$,
>
> $\qquad$ variance $= \sigma^2 = \text{Var}(X) = \lambda$.
>
> The mean and variance of a Poisson distribution are equal.

The equality of the mean and variance of a Poisson distribution gives a simple way of testing whether a variable might be modelled by a Poisson distribution. The mean of the data in Table 1.2 has already been used and is equal to 0.37. You can verify that the variance of these data is 0.4331. These values, which are both 0.4 to 1 decimal place, are sufficiently close to indicate that the Poisson distribution may be a suitable model for the number of phone calls in a 5-minute interval. This is confirmed by Table 1.4, which shows that the relative frequencies calculated from Table 1.2 are close to the theoretical probabilities found by assuming that $X \sim \text{Po}(0.37)$. (The values for the probabilities are given to 3 decimal places and the value for $\text{P}(X \geqslant 4)$ has been found by subtraction.)

Note that if the mean and variance are not approximately equal then the Poisson distribution is not a suitable model. If they are equal then the Poisson distribution may be a suitable model, but is not necessarily so.

x	Frequency	Relative frequency	$P(X = x)$
0	71	0.71	$e^{-0.37} = 0.691$
1	23	0.23	$e^{-0.37}0.37 = 0.256$
2	4	0.04	$e^{-0.37}\dfrac{0.37^2}{2!} = 0.047$
3	2	0.02	$e^{-0.37}\dfrac{0.37^3}{3!} = 0.006$
$\geqslant 4$	0	0	0
Totals	100	1	1

Table 1.4. Comparison of theoretical Poisson probabilities and relative frequencies for the data in Table 1.2.

Exercise 1B

1 For each of the following situations, say whether or not the Poisson distribution might provide a suitable model.

(a) The number of raindrops that fall onto an area of ground of 1 cm^2 in a period of 1 minute during a shower.

(b) The number of occupants of vehicles that pass a given point on a busy road in 1 minute.

(c) The number of flaws in a given length of material of constant width.

(d) The number of claims made to an insurance company in a month.

2 Weeds grow on a large lawn at an average rate of 5 per square metre. A particular metre square is considered and sub-divided into smaller and smaller squares. Copy and complete the table below, assuming that no more than 1 weed can grow in a sub-division.

Number of sub-divisions	P(a sub-division contains a weed)	P(no weeds in a given square metre)
100	$\frac{5}{100} = 0.05$	$0.95^{100} = 0.005\,921$
10 000	$\frac{5}{10\,000} =$	
1 000 000		
100 000 000		

Compare your answers to the probability of no weeds in a given square metre, given by the Poisson probability formula.

3 The number of telephone calls I received during the month of March is summarised in the table.

Number of telephone phone calls received per day (x)	0	1	2	3	4
Number of days	9	12	5	4	1

(a) Calculate the relative frequency for each of $x = 0, 1, 2, 3, 4$.

(b) Calculate the mean and variance of the distribution. (Give your answers correct to 2 decimal places.) Comment on the suitability of the Poisson distribution as a model for this situation.

(c) Use the Poisson distribution to calculate $P(X = x)$, for $x = 0, 1, 2, 3$ and ≥ 4 using the mean calculated in part (b).

(d) Compare the theoretical probabilities and the relative frequencies found in part (a). Do these figures support the comment made in part (b)?

4 The number of goals scored by a football team during a season gave the following results.

Number of goals per match	0	1	2	3	4	5	6	7
Number of matches	5	19	9	5	2	1	0	1

Calculate the mean and variance of the distribution. Calculate also the relative frequencies and theoretical probabilities for $x = 0, 1, 2, 3, 4, 5, 6, \geq 7$, assuming a Poisson distribution with the same mean. Do you think, in the light of your calculations, that the Poisson distribution provides a suitable model for the number of goals scored per match?

5 The number of cars passing a given point in 100 10-second intervals was observed as follows.

Number of cars	0	1	2	3	4	5
Number of intervals	47	33	16	3	0	1

Do you think that a Poisson distribution is a suitable model for these data?

1.4 Practical activities

1 Traffic flow   In order to carry out this activity you will need to make your observations on a road where the traffic flows freely, preferably away from traffic lights, junctions etc. The best results will be obtained if the rate of flow is one to two cars per minute on average.
(a) Count the number of cars which pass each minute over a period of one hour and assemble your results into a frequency table.
(b) Calculate the mean and variance of the number of cars per minute. Comment on your results.
(c) Compare the relative frequencies with the Poisson probabilities calculated by taking λ equal to the mean of your data. Comment on the agreement between the two sets of values.

2 Random rice   For this activity you need a chessboard and a few tablespoonfuls of uncooked rice.

(a) Scatter the rice 'at random' on to the chessboard. This can be achieved by holding your hand about 50 cm above the board and moving it around as you drop the rice. Drop sufficient rice to result in two to three grains of rice per square on average.

(b) Count the number of grains of rice in each square and assemble your results into a frequency table.

(c) Calculate the mean and variance of the number of grains per square. If these are reasonably close then go on to part (d). If not, see if you can improve your technique for scattering rice 'at random'!

(d) Compare the relative frequencies with the Poisson probabilities calculated taking λ equal to the mean of your data. Comment on the agreement between the two sets of values.

3 Background radiation   For this activity you need a Geiger counter with a digital display. When the Geiger counter is switched on it will record the background radiation.

(a) Prepare a table in which you can record the reading on the Geiger counter every 5 seconds for total time of 5 minutes.

(b) Switch the counter on and record the reading every 5 seconds.

(c) Plot a graph of the reading on the counter against time taking values every 30 seconds. Does this graph suggest that the background rate is constant?

(d) The number of counts in each 5 second interval can be found by taking the difference between successive values in the table which you made in parts (a) and (b). Find these values and assemble them into a frequency table.

(e) Calculate the mean and variance of the number of counts per 5 seconds. Comment on your results.

(f) Compare the relative frequencies with the Poisson probabilities calculated by taking λ equal to the mean of your data. Comment on the agreement between the two sets of values.

4 Football goals   For this activity you need details of the results of the matches in a football division for one particular week.

(a) Make a frequency table of the number of goals scored by each team.

(b) Calculate the mean and variance of the number of goals scored.

(c) Compare the relative frequencies with the Poisson probabilities calculated by taking λ equal to the mean of your data.

(d) Discuss whether the variable 'number of goals scored by each team' satisfies the conditions required for the Poisson distribution to be a suitable model. Comment on the results you obtained in part (b) and part (c) in the light of your answer.

1.5   The Poisson distribution as an approximation to the binomial distribution

In certain circumstances it is possible to use the Poisson distribution rather than the binomial distribution in order to make the calculation of probabilities easier.

Consider items coming off a production line. Suppose that some of the items are defective and that defective items occur at random with a constant probability of 0.03. The items are packed in boxes of 200 and you want to find the probability that a box contains two or fewer defective items.

The number, X, of defective items in box has a binomial distribution since there are

- a fixed number (200) of items in each box,
- each item is either defective or not,
- the probability of a defective item is constant and equal to 0.03,
- defective items occur independently of each other.

This means that $X \sim B(200, 0.03)$. The probability that a box contains two or fewer defective items can be calculated exactly using the binomial distribution as follows.

$$P(X \leqslant 2) = P(X = 0) + P(X = 1) + P(X = 2)$$
$$= 0.97^{200} + \binom{200}{1} 0.97^{199} 0.03 + \binom{200}{2} 0.97^{198} 0.03^2$$
$$= 0.002\ 261 \ldots + 0.013\ 987 \ldots + 0.043\ 042 \ldots$$
$$= 0.0592, \text{ correct to 3 significant figures.}$$

This binomial distribution has a large value of n and a small value of p. This is exactly the situation which applied in Section 1.2 when the Poisson distribution was treated as a limiting case of the binomial distribution. In these circumstances, that is large n and small p, the probabilities can be calculated approximately using a Poisson distribution whose mean is equal to the mean of the binomial distribution. The mean of the binomial distribution is given by $np = 200 \times 0.03 = 6$ (see S1 Section 8.3). Using $X \sim Po(6)$ gives for the required probability

$$P(X \leqslant 2) = P(X = 0) + P(X = 1) + P(X = 2)$$
$$= e^{-6} + e^{-6}6 + e^{-6}\frac{6^2}{2!}$$
$$= 0.002\ 478 \ldots + 0.014\ 872 \ldots + 0.044\ 617 \ldots$$
$$= 0.0620, \text{ correct to 3 significant figures.}$$

If you follow though the calculations using a calculator you will find that the calculation using the Poisson distribution is much easier to perform. Using the Poisson distribution only gives an approximate answer. In this case the answers for the individual probabilities and the value for $P(X \leqslant 2)$ agree to 1 significant figure. This is often good enough for practical purposes.

It is important to remember that the approximate method using the Poisson distribution will only give reasonable agreement with the exact method using the binomial distribution when n is large and p is small. The larger n and the smaller p, the better the agreement between the two answers. In practice you should not use the approximate method unless n is large and p is small. A useful rule of thumb is that $n > 50$ and $np < 5$.

> If $X \sim B(n, p)$, and if $n > 50$ and $np < 5$, then X can reasonably be approximated by the Poisson distribution $W \sim Po(np)$.
> The larger n and the smaller p, the better the approximation.

Example 1.5.1

Calculate the following probabilities, using a suitable approximation where appropriate.

(a) $P(X < 3)$ given that $X \sim B(100, 0.02)$.
(b) $P(X < 10)$ given that $X \sim B(60, 0.3)$.
(c) $P(X < 2)$ given that $X \sim B(10, 0.01)$.

(a) Here n is large (that is greater than 50) and p is small, which suggests that a Poisson approximation may be appropriate. As a check calculate $np = 100 \times 0.02 = 2$. Since $np < 5$ the Poisson approximation, $W \sim Po(2)$, may be used. Using the Poisson formula

$$P(W < 3) = P(W \leqslant 2) = P(W = 0) + P(W = 1) + P(W = 2)$$

$$= e^{-2} + e^{-2}2 + e^{-2}\frac{2^2}{2!}$$

$$= 0.677, \text{ correct to 3 significant figures.}$$

(b) Here n is still large (that is greater than 50) but p is not small enough to make np $(= 60 \times 0.3 = 18)$ less than 5. However np $(= 18)$ and nq $(= 60 \times 0.7 = 42)$ are both greater than 5 so the normal approximation to the binomial distribution, which you met in S1 Section 9.7, may be used. The mean of the binomial distribution is 18 and the variance is $npq = 60 \times 0.3 \times 0.7 = 12.6$, so $X \sim B(60, 0.3)$ is approximated by $V \sim N(18, 12.6)$ with a continuity correction.

$$P(X < 10) = P(V \leqslant 9.5) = P\left(Z \leqslant \frac{9.5 - 18}{\sqrt{12.6}}\right) = P(Z \leqslant -2.395)$$

$$= 1 - \Phi(2.395)$$

$$= 1 - 0.9917 \quad \text{(using the table on page 165)}$$

$$= 0.008, \text{ correct to 3 decimal places.}$$

(c) Here p is small but n is not large enough to use the Poisson approximation. The normal approximation should not be used either since $np = 10 \times 0.01 = 0.1$ is not greater than 5. In fact it is not appropriate to use an approximation at all. The required probability must be calculated using the binomial probability formula as follows.

$$P(X < 2) = P(X = 0) + P(X = 1)$$

$$= \binom{10}{0}0.99^{10}0.01^0 + \binom{10}{1}0.99^9 0.01^1$$

$$= 0.9043\ldots + 0.091\,35\ldots$$

$$= 0.996, \text{ correct to 3 significant figures.}$$

Exercise 1C

1 (a) There are 1000 pupils in a school. Find the probability that exactly 3 of them have their birthdays on 1 January, by using

    (i) $B\left(1000, \frac{1}{365}\right)$,   (ii) $Po\left(\frac{1000}{365}\right)$.

  (b) There are 5000 students in a university. Calculate the probability that exactly 15 of them have their birthdays on 1 January, by using

    (i) a suitable binomial distribution,   (ii) a suitable Poisson approximation.

For the rest of the exercise, use, where appropriate, the Poisson approximation to the binomial distribution.

2 If $X \sim B(300, 0.004)$ find

  (a) $P(X < 3)$,   (b) $P(X > 4)$.

3 The probability that a patient has a particular disease is 0.008. One day 80 people go to their doctor.

  (a) What is the probability that exactly 2 of them have the disease?

  (b) What is the probability that 3 or more of them have the disease?

4 The probability of success in an experiment is 0.01. Find the probability of 4 or more successes in 100 trials of the experiment.

5 When eggs are packed in boxes the probability that an egg is broken is 0.008.

  (a) What is the probability that in a box of 6 eggs there are no broken eggs?

  (b) Calculate the probability that in a consignment of 500 eggs fewer than 4 eggs are broken.

6 When a large number of flashlights leaving a factory is inspected it is found that the bulb is faulty in 1% of the flashlights and the switch is faulty in 1.5% of them. Assuming that the faults occur independently and at random, find

  (a) the probability that a sample of 10 flashlights contains no flashlights with a faulty bulb,

  (b) the probability that a sample of 80 flashlights contains at least one flashlight with both a defective bulb and a defective switch,

  (c) the probability that a sample of 80 flashlights contains more than two defective flashlights.

1.6 The normal distribution as an approximation to the Poisson distribution

Example 1.5.1(b) gave a reminder of the method for using the normal distribution as an approximation to the binomial distribution. The normal distribution may be used in a similar way as an approximation to the Poisson distribution provided that the mean of the Poisson distribution is sufficiently large.

Fig. 1.5 shows why such an approximation is valid: as the value of λ increases, the shape of the Poisson distribution becomes more like the characteristic bell shape of the normal distribution.

If you have access to a computer, you can use a spreadsheet to draw these diagrams for yourself.

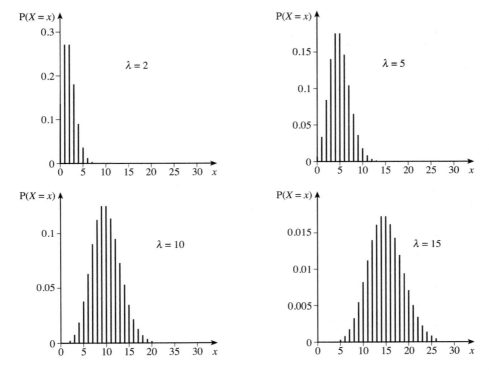

Fig. 1.5. Bar charts showing the Poisson distribution for different values of λ.

Since the variance of a Poisson distribution is equal to its mean, both the mean and variance of the normal distribution which is used as an approximation are taken to be equal to λ. Just as for the normal approximation to the binomial distribution, a continuity correction is needed because a discrete distribution is being approximated by a continuous one. As a rule-of-thumb the normal approximation to the Poisson distribution should only be used if $\lambda > 15$. You can see from the last diagram in Fig. 1.5 that this looks very reasonable.

> If $X \sim \text{Po}(\lambda)$ and if $\lambda > 15$ then X may reasonably be approximated by the normal distribution $Y \sim \text{N}(\lambda, \lambda)$.
>
> A continuity correction must be applied.
>
> The larger λ the better the approximation.

Example 1.6.1

It is thought that the number of serious accidents, X, in a time interval of t weeks, on a given stretch of road, can be modelled by a Poisson distribution with mean $0.4t$. Find the probability of

(a) one or fewer accidents in a randomly chosen 2-week interval,

(b) 12 or more accidents in a randomly chosen year.

(a) For a time interval of two weeks, $\lambda = 0.4 \times 2 = 0.8$.

$$P(X \leqslant 1) = P(X = 0) + P(X = 1)$$
$$= e^{-0.8} + e^{-0.8}0.8$$
$$= 0.809, \text{correct to 3 decimal places.}$$

Note that, since $\lambda \leqslant 15$, the normal approximation is not appropriate.

(b) For a time interval of 1 year, $\lambda = 0.4 \times 52 = 20.8$.

Since $\lambda > 15$, a normal approximation is appropriate. $X \sim \text{Po}(20.8)$ is approximated by $Y \sim N(20.8, 20.8)$, with a continuity correction.

$$P(X \geqslant 12) = P(Y > 11.5) = P\left(Z > \frac{11.5 - 20.8}{\sqrt{20.8}}\right) = P(Z > -2.039)$$
$$= P(Z \leqslant 2.039) = \Phi(2.039)$$
$$= 0.9792 = 0.979, \text{correct to 3 decimal places.}$$

Exercise 1D

Use the normal approximation to the Poisson distribution, where appropriate.

1   If $X \sim \text{Po}(30)$ find

    (a) $P(X \leqslant 31)$,   (b) $P(35 \leqslant X \leqslant 40)$,   (c) $P(29 < X \leqslant 32)$.

2   Accidents occur in a factory at an average rate of 5 per month. Find the probabilities that

    (a) there will be fewer than 4 in a month,   (b) there will be exactly 62 in a year.

3   The number of accidents on a road follows a Poisson distribution with a mean of 8 per week. Find the probability that in a year (assumed to be 52 weeks) there will be fewer than 400 accidents.

4   Insect larvae are distributed at random in a pond at a mean rate of 8 per m^3 of pond water. The pond has a volume of 40 m^3. Calculate the probability that there are more than 350 insect larvae in the pond.

5   Water taken from a river contains on average 16 bacteria per ml. Assuming a Poisson distribution find the probability that 5 ml of the water contains

    (a) from 65 to 85 bacteria, inclusive,   (b) exactly 80 bacteria.

6   A company receives an average of 40 telephone calls an hour. The number of calls follows a Poisson distribution.

(a) Find the probability that there are from 35 to 50 calls (inclusive) in a given hour.

(b) Find the probability that there are exactly 42 calls in a given hour.

7   Given that $X \sim \text{Po}(50)$ and $\text{P}(X > x) \leqslant 0.05$, find the minimum integer value of x.

8   Sales of cooking oil bought in a shop during a week follow a Poisson distribution with mean 100. How many units should be kept in stock to be at least 99% certain that supply will be able to meet demand?

Miscellaneous exercise 1

1   Between the hours of 0800 and 2200, cars arrive at a certain petrol station at an average rate of 0.8 per minute. Assuming that arrival times are random, calculate the probability that at least 2 cars will arrive during a particular minute between 0800 and 2200.   (OCR)

2   The proportion of patients who suffer an allergic reaction to a certain drug used to treat a particular medical condition is assumed to be 0.045.

Each of a random sample of 90 patients with the condition is given the drug and X is the number who suffer an allergic reaction. Assuming independence, explain why X can be modelled approximately by a Poisson distribution and calculate $\text{P}(X = 4)$.   (OCR)

3   The number of night calls to a fire station in a small town can be modelled by a Poisson distribution with mean 4.2 per night. Find the probability that on a particular night there will be 3 or more calls to the fire station.

State what needs to be assumed about the calls to the fire station in order to justify a Poisson model.   (OCR)

4   On average, a cycle shop sells 1.8 cycles per week. Assuming that the sales occur at random,

(a) find the probability that exactly 2 cycles are sold in a given week,

(b) find the probability that exactly 4 cycles are sold in a given two-week period.

5   A householder wishes to sow part of her garden with grass seed. She scatters seed randomly so that the number of seeds falling on any particular region is a random variable having a Poisson distribution, with its mean proportional to the area of the region. The part of the garden that she intends to sow has area 50 m^2 and she estimates that she will sow 10^6 seeds. Calculate the expected number of seeds falling on a region, R, of area 1 cm^2, and show that the probability that no seeds fall on R is 0.135, correct to 3 significant figures.

The number of seeds falling on R is denoted by X. Find the probability that either $X = 0$ or $X > 4$.

The number of seeds falling on a region of area 100 cm^2 is denoted by Y. Using a normal approximation, find $\text{P}(175 \leqslant Y \leqslant 225)$.   (OCR)

6 It is given that 93% of children in the UK have been immunised against whooping cough. The number of children in a random sample of 60 children who have been immunised is X, and the number not immunised is Y. State, with justification, which of X or Y has a distribution which can be approximated by a Poisson distribution.

Using a Poisson approximation, estimate the probability that at least 58 children from the sample have been immunised against whooping cough.   (OCR)

7 A firm investigated the number of employees suffering injuries whilst at work. The results recorded below were obtained for a 52-week period.

Number of employees injured in a week	0	1	2	3	4 or more
Number of weeks	31	17	3	1	0

Give reasons why one might expect this distribution to approximate to a Poisson distribution. Evaluate the mean and variance of the data and explain why this gives further evidence in favour of a Poisson distribution.

Using the calculated value of the mean, find the theoretical frequencies of a Poisson distribution for the number of weeks in which 0, 1, 2, 3, 4 or more employees were injured.   (OCR)

8 Analysis of the scores in football matches in a local league suggests that the total number of goals scored in a randomly chosen match may be modelled by the Poisson distribution with parameter 2.7. The number of goals scored in different matches are independent of one another.

(a) Find the probability that a match will end with no goals scored.

(b) Find the probability that 4 or more goals will be scored in a match.

One Saturday afternoon, 11 matches are played in the league.

(c) State the expected number of matches in which no goals are scored.

(d) Find the probability that there are goals scored in all 11 matches.

(e) State the distribution for the total number of goals scored in the 11 matches. Using a suitable approximating distribution, or otherwise, find the probability that more than 30 goals are scored in total.   (MEI)

9 The discrete random variable X has probability distribution as shown in the table below, where p is a constant.

x	0	1	2	3
$P(X = x)$	p	$\frac{1}{2}p$	$\frac{1}{4}p$	$\frac{1}{20}p$

Show that $p = \frac{5}{9}$.

One hundred independent observations of X are made, and the random variable Y denotes the number of occasions on which $X = 3$. Explain briefly why the distribution of Y may be approximated by a suitable Poisson distribution, and state the mean of this Poisson distribution.

Find $P(Y = 2)$ and $P(Y \geq 4)$, giving your answers to 3 significant figures.   (OCR)

10 Data files on computers have sizes measured in megabytes. When files are sent from one computer to another down a communications link, the number of errors has a Poisson distribution. On average, there is one error for every 10 megabytes of data.

(a) Find the probability that a 3 megabyte file is transmitted

(i) without error, (ii) with 2 or more errors.

(b) Show that a file which has a 95% chance of being transmitted without error is a little over half a megabyte in size.

A commercial organisation transmits 1000 megabytes of data per day.

(c) State how many errors per day they will incur on average.

Using a suitable approximating distribution, show that the number of errors on any randomly chosen day is virtually certain to be between 70 and 130. (MEI)

11 A manufacturer produces an integrated electronic unit which contains 36 separate pressure sensors. Due to difficulties in manufacture, it happens very often that not all the sensors in a unit are operational. 100 units are tested and the number N of pressure sensors which function correctly are distributed according to the table.

N	36	35	34	33	32	31	30	29	28	≤ 27
Number of units	5	15	22	22	17	11	5	2	1	0

Calculate the mean number of sensors which are faulty.

The manufacturer only markets those units which have at least 32 of their 36 sensors operational. Estimate, using the Poisson distribution, the percentage of units which are not marketed. (OCR)

12 An aircraft has 116 seats. The airline has found, from long experience, that on average 2.5% of people with tickets for a particular flight do not arrive for that flight. If the airline sells 120 tickets for a particular flight determine, using a suitable approximation, the probability that more than 116 people arrive for that flight. Determine also the probability that there are empty seats on the flight. (OCR)

2 Linear combinations of random variables

This chapter studies the distribution of linear functions and linear combinations of random variables. When you have completed it you should be able to apply the following results.

- $E(aX+b) = aE(X)+b$ and $\mathrm{Var}(aX+b) = a^2\mathrm{Var}(X)$
- $E(aX+bY) = aE(X)+bE(Y)$
- $\mathrm{Var}(aX+bY) = a^2\mathrm{Var}(X)+b^2\mathrm{Var}(Y)$ for independent X and Y
- if X has a normal distribution, then so does $aX+b$
- if X and Y have independent normal distributions, then $aX+bY$ has a normal distribution
- if X and Y have independent Poisson distributions, then $X+Y$ has a Poisson distribution.

2.1 The expectation and variance of a linear function of a random variable

In S1 Chapter 8 you learnt how to calculate the mean and variance of a discrete random variable. Here is an example.

The probability distribution of a discrete random variable, X, is

x	1	2	3	4
$P(X = x)$	0.1	0.2	0.3	0.4

giving

$$\mu_X = E(X) = \sum xP(X = x)$$
$$= 1 \times 0.1 + 2 \times 0.2 + 3 \times 0.3 + 4 \times 0.4$$
$$= 0.1 + 0.4 + 0.9 + 1.6$$
$$= 3,$$

and

$$\mathrm{Var}(X) = \sum x^2 P(X = x) - \mu_X{}^2$$
$$= 1^2 \times 0.1 + 2^2 \times 0.2 + 3^2 \times 0.3 + 4^2 \times 0.4 - 3^2$$
$$= 0.1 + 0.8 + 2.7 + 6.4 - 9$$
$$= 1.$$

Suppose that you now define a new random variable Y such that $Y = X + 4$. Would there be any relationship between the mean and variance of Y and the mean and variance of X? Maybe you can guess the values of $E(Y)$ and $\mathrm{Var}(Y)$. You can check your prediction by calculating $E(Y)$ and $\mathrm{Var}(Y)$.

The probability distribution of Y is

y	5	6	7	8
$P(Y = y)$	0.1	0.2	0.3	0.4

giving

$$\mu_Y = E(Y) = \sum yP(Y = y)$$
$$= 5 \times 0.1 + 6 \times 0.2 + 7 \times 0.3 + 8 \times 0.4$$
$$= 0.5 + 1.2 + 2.1 + 3.2$$
$$= 7,$$

and

$$\mathrm{Var}\,(Y) = \sum y^2 P(Y = y) - \mu_Y{}^2$$
$$= 5^2 \times 0.1 + 6^2 \times 0.2 + 7^2 \times 0.3 + 8^2 \times 0.4 - 7^2$$
$$= 2.5 + 7.2 + 14.7 + 25.6 - 49$$
$$= 1.$$

You can see that $E(Y) = E(X) + 4$ and $\mathrm{Var}\,(Y) = \mathrm{Var}\,(X)$.

These are examples of general results.

If $Y = X + b$, where b is a constant, then

$$E(Y) = E(X + b) = E(X) + b \qquad (2.1)$$

and

$$\mathrm{Var}\,(Y) = \mathrm{Var}\,(X + b) = \mathrm{Var}\,(X). \qquad (2.2)$$

You may be surprised that the variance is unchanged. However, this result is reasonable when you remember that variance is a measure of spread and that adding a constant to X (or subtracting a constant from it) does not affect the spread of values.

What would be the effect of multiplying X by a constant? Suppose, for example, you defined a new variable W such that $W = 5X$. Then the probability distribution of W is given by the table.

w	5	10	15	20
$P(W = w)$	0.1	0.2	0.3	0.4

It is left as an exercise for you to show that $E(W) = 15$ and $\mathrm{Var}\,(W) = 25$. Note that $E(W) = 5E(X)$ and $\mathrm{Var}\,(W) = 5^2 \mathrm{Var}\,(X)$.

These are examples of other general results.

If $W = aX$, where a is a constant, then

$$E(W) = E(aX) = aE(X) \qquad (2.3)$$

and

$$\text{Var}(W) = \text{Var}(aX) = a^2\text{Var}(X). \qquad (2.4)$$

These four results can be combined to find the mean and variance of a linear function of X of the form $aX + b$, where a and b are constants, as follows.

$$\begin{aligned} E(aX + b) &= E(aX) + b & \text{from (2.1)} \\ &= aE(X) + b & \text{from (2.3),} \end{aligned}$$

and

$$\begin{aligned} \text{Var}(aX + b) &= \text{Var}(aX) & \text{from (2.2)} \\ &= a^2\text{Var}(X) & \text{from (2.4).} \end{aligned}$$

Here are proofs of these results in the case when X is a discrete random variable.

Theorem   For a discrete random variable, $E(aX + b) = aE(X) + b$.

Proof   Let the discrete random variable X take values x_i with probabilities p_i. Then,

$$E(X) = \mu_X = \sum x_i p_i \;.$$

Now define a new variable Y such that $Y = aX + b$. This variable Y takes values y_i, where $y_i = ax_i + b$, with probabilities p_i.

The expected value of Y is given by

$$\begin{aligned} E(Y) = \mu_Y &= \sum y_i p_i = \sum (ax_i + b) p_i \\ &= a\sum x_i p_i + b\sum p_i \\ &= aE(X) + b & (\text{since } \sum p_i = 1). \end{aligned}$$

Since $Y = aX + b$, $E(Y) = E(aX + b)$; it follows that $E(aX + b) = aE(X) + b$.

Theorem   For a discrete random variable, $\text{Var}(aX + b) = a^2\text{Var}(X)$.

Proof   From S1 Section 8.2,

$$\text{Var}(X) = \sum (x_i - \mu_X)^2 p_i = \sum x_i^2 p_i - \mu_X^2 \;.$$

The first form of the formula for variance is more convenient for this proof.

Now define a new variable Y such that $Y = aX + b$. This variable Y takes values y_i, where $y_i = ax_i + b$, with probabilities p_i.

For the variable Y

$$
\begin{aligned}
\text{Var}(Y) &= \sum (y_i - \mu_Y)^2 p_i \\
&= \sum ((ax_i + b) - (a\mu_X + b))^2 p_i \\
&= \sum (a(x_i - \mu_X))^2 p_i \\
&= \sum a^2 (x_i - \mu_X)^2 p_i = a^2 \sum (x_i - \mu_X)^2 p_i \\
&= a^2 \text{Var}(X).
\end{aligned}
$$

Since $Y = aX + b$, $\text{Var}(Y) = \text{Var}(aX + b)$; it follows that $\text{Var}(aX + b) = a^2 \text{Var}(X)$.

These results also hold for continuous random variables. (Chapter 3 discusses random variables in detail.)

Example 2.1.1

The temperature in degrees Fahrenheit on a remote island is a random variable with mean 59 and variance 27. Find the mean and variance of the temperature in degrees Centigrade, given that to convert degrees Fahrenheit to degrees Centigrade you subtract 32 and then multiply by $\frac{5}{9}$.

Let X be the temperature in °F. Then $\text{E}(X) = 59$ and $\text{Var}(X) = 27$.

If Y is the temperature in °C, then $Y = \frac{5}{9}(X - 32) = \frac{5}{9}X - \frac{160}{9}$.

So $\quad \text{E}(Y) = \text{E}\left(\frac{5}{9}X - \frac{160}{9}\right) = \frac{5}{9}\text{E}(X) - \frac{160}{9} = \frac{5}{9} \times 59 - \frac{160}{9} = 15$,

and $\quad \text{Var}(Y) = \left(\frac{5}{9}\right)^2 \text{Var}(X) = \frac{25}{81} \times 27 = \frac{25}{3}$.

For any random variable X,

$$\text{E}(aX + b) = a\text{E}(X) + b, \qquad (2.5)$$

$$\text{Var}(aX + b) = a^2 \text{Var}(X), \qquad (2.6)$$

where a and b are constants.

2.2 Linear combinations of more than one random variable

Situations often arise in which you know the expected value and variance of each of several random variables and would like to find the expected value and variance of a linear combination. For example, you might know the expected value and variance of the thickness of the sheets which make up a laminated windscreen and wish to find the expected value and variance of the total thickness of the windscreen.

In order to investigate any possible relations between expected values and variances you could start by considering two discrete random variables, X and Y. Table 2.1 gives the probability distributions of two such random variables.

x	1	2	3
$P(X = x)$	$\frac{1}{2}$	$\frac{1}{3}$	$\frac{1}{6}$

y	2	4
$P(Y = y)$	$\frac{1}{3}$	$\frac{2}{3}$

Table 2.1. Probability distributions of two discrete random variables.

It is left as an exercise for you to show that $E(X) = \frac{5}{3}$, $Var(X) = \frac{5}{9}$, $E(Y) = \frac{10}{3}$ and $Var(Y) = \frac{8}{9}$.

The possible values of S, where $S = X + Y$ are given below.

		Value of X		
		1	2	3
Value of Y	2	3	4	5
	4	5	6	7

The corresponding probabilities for S are given below.

		Value of X		
		1	2	3
Value of Y	2	$\frac{1}{2} \times \frac{1}{3}$	$\frac{1}{3} \times \frac{1}{3}$	$\frac{1}{6} \times \frac{1}{3}$
	4	$\frac{1}{2} \times \frac{2}{3}$	$\frac{1}{3} \times \frac{2}{3}$	$\frac{1}{6} \times \frac{2}{3}$

Assembling these results into a probability distribution for S gives Table 2.2. The fractions have been given the same denominator in order to show the shape of the distribution.

s	3	4	5	6	7
$P(S = s)$	$\frac{3}{18}$	$\frac{2}{18}$	$\frac{7}{18}$	$\frac{4}{18}$	$\frac{2}{18}$

Table 2.2. Probability distribution of S where $S = X + Y$.

Now check that $E(S) = 5$ and $Var(S) = \frac{13}{9}$. You should be able to spot that the relation between the expected values is $E(S) = E(X) + E(Y)$ and between the variances is $Var(S) = Var(X) + Var(Y)$. These are particular instances of general results which hold for both discrete and continuous random variables. However, it is important to appreciate that while the relation between expected values is true for *all* random variables the relation between the variances is only true for *independent* random variables.

The relations $E(S) = E(X) + E(Y)$ and $Var(S) = Var(X) + Var(Y)$ can be extended to situations in which there are more than two random variables.

Example 2.2.1

My journey to work is made up of four stages: a walk to the bus-stop, a wait for the bus, a bus journey and a walk at the other end. The times taken for these four stages are independent random variables U, V, W and X with expected values (in minutes) of $4.7, 5.6, 21.6$ and 3.7 respectively and standard deviations of $1.1, 1.2, 3.1$ and 0.8 respectively. What is the expected time and standard deviation for the total journey?

The expected time for the whole journey is

$$E(U + V + W + X) = E(U) + E(V) + E(W) + E(X)$$
$$= 4.7 + 5.6 + 21.6 + 3.7 = 35.6.$$

Since the variables U, V, W and X are independent

$$Var(U + V + W + X) = Var(U) + Var(V) + Var(W) + Var(X)$$
$$= 1.1^2 + 1.2^2 + 3.1^2 + 0.8^2 = 12.9,$$

so the standard deviation for the whole journey is $\sqrt{12.9} = 3.59$, correct to 3 significant figures.

The relations between expected values and variances can also be generalised to the situation where $S = aX + bY$ and a and b are constants. Then

$$E(S) = E(aX + bY) = E(aX) + E(bY)$$
$$= aE(X) + bE(Y) \qquad \text{(using Equation 2.5),}$$

and for independent X and Y

$$Var(S) = Var(aX + bY) = Var(aX) + Var(bY)$$
$$= a^2 Var(X) + b^2 Var(Y) \qquad \text{(using Equation 2.6).}$$

Example 2.2.2

The length, L (in cm), of the boxes produced by a machine is a random variable with mean 26 and variance 4 and the width, B (in cm), is a random variable with mean 14 and variance 1. The variables L and B are independent. What are the expected value and variance of
(a) the perimeter of the boxes,
(b) the difference between the length and the width?

(a) The perimeter is $2L + 2B$, so

$$E(2L + 2B) = 2E(L) + 2E(B) = 2 \times 26 + 2 \times 14 = 80,$$
$$Var(2L + 2B) = 2^2 Var(L) + 2^2 Var(B) = 4 \times 4 + 4 \times 1 = 20.$$

(b) The difference between length and width is $L - B$, so

$$E(L - B) = E(L) - E(B) = 26 - 14 = 12,$$
$$Var(L - B) = 1^2 Var(L) + (-1)^2 Var(B) = 1 \times 4 + 1 \times 1 = 5.$$

The result for the variance is an example of a general rule which applies when variances are combined, namely that they are always added.

For any random variables X, Y, U, V etc.,

$$E(aX + bY + cU + dV + \ldots) = aE(X) + bE(Y) + cE(U) + dE(V) + \ldots . \quad (2.7)$$

For *independent* random variables X, Y, U, V etc.,

$$\mathrm{Var}(aX + bY + cU + dV + \ldots)$$
$$= a^2 \mathrm{Var}(X) + b^2 \mathrm{Var}(Y) + c^2 \mathrm{Var}(U) + d^2 \mathrm{Var}(V) + \ldots . \quad (2.8)$$

2.3 Linear relations involving more than one observation of a random variable

Equations 2.7 and 2.8 can be applied to repeated observations of a single random variable. Suppose, for example, you make two observations of a random variable X where $E(X) = 3$ and $\mathrm{Var}(X) = 4$. Denote these observations by X_1 and X_2. Then the expected values and variances of X_1 and X_2 will be equal to the corresponding values for X. It follows that

$$E(X_1 + X_2) = E(X_1) + E(X_2) = E(X) + E(X) = 3 + 3 = 6,$$

and, providing the observations are independent, that

$$\mathrm{Var}(X_1 + X_2) = \mathrm{Var}(X_1) + \mathrm{Var}(X_2) = \mathrm{Var}(X) + \mathrm{Var}(X) = 4 + 4 = 8.$$

It is instructive to compare these results with the values for $E(2X)$ and $\mathrm{Var}(2X)$ obtained using Equations 2.5 and 2.6:

$$E(2X) = 2E(X) = 2 \times 3 = 6, \quad \text{and} \quad \mathrm{Var}(2X) = 2^2 \mathrm{Var}(X) = 2^2 \times 4 = 16.$$

The expected values for $X_1 + X_2$ and $2X$ are the same. This is not surprising since $2X = X + X$ and so $E(2X) = E(X) + E(X) = 6$. Why then do the variances of $X_1 + X_2$ and $2X$ differ? The answer is that it is not true to say that $\mathrm{Var}(2X) = \mathrm{Var}(X) + \mathrm{Var}(X)$, because the variables X and X are not independent: they refer to the *same* observation of the same variable. This means that it is important to distinguish clearly between situations in which a single observation is multiplied by a constant and those in which several different observations of the same random variable are added.

Exercise 2A

1 The random variable X has the probability distribution shown below.

x	0	1	2
$P(X = x)$	0.2	0.3	0.5

(a) Find $E(X)$ and $\mathrm{Var}(X)$.

A variable Y is defined by $Y = 3X + 2$.

(b) Use Equations 2.5 and 2.6 to calculate $E(Y)$ and $\mathrm{Var}(Y)$.

(c) Verify your answers to part (b) by calculating $E(Y)$ and $\mathrm{Var}(Y)$ from the probability distribution of Y.

2 A random variable X has mean 24 and variance 5. Find the mean and variance of

(a) $20 - X$,   (b) $4X - 7$.

3 Three discs are drawn at random from a box containing a number of red and blue discs. The mean and variance of the number of blue discs drawn are 2 and 0.5 respectively. What are the mean and variance of the number of red discs drawn?

4 The random variable X is the number of even numbers obtained when two ordinary fair dice are thrown. The random variable Y is the number of even numbers obtained when two fair pentagonal spinners, each numbered 1, 2, 3, 4, 5, are spun simultaneously.

Copy and complete the following probability distributions.

x	0	1	2
p	0.25		

y	0	1	2
p	0.36		

$x + y$	0	1	2	3	4
p	0.09				

Find $E(X)$, $Var(X)$, $E(Y)$, $Var(Y)$, $E(X+Y)$, $Var(X+Y)$ by using the probability distributions.

Verify that $E(X+Y) = E(X) + E(Y)$ and $Var(X+Y) = Var(X) + Var(Y)$.

5 X and Y are independent random variables with probability distributions as shown.

x	1	2	3
p	0.4	0.2	0.4

y	0	1	2
p	0.3	0.5	0.2

You are given that $E(X) = 2$, $Var(X) = 0.8$, $E(Y) = 0.9$ and $Var(Y) = 0.49$.

The random variable T is defined as $2X - Y$. Find $E(T)$ and $Var(T)$ using Equations 2.7 and 2.8.

Check your answers by completing the following probability distribution and calculating $E(T)$ and $Var(T)$ directly.

t	0	1	2	3	4	5	6
p	0.08						0.12

6 The independent random variables W, X and Y have means 10, 8 and 6 respectively and variances 4, 5 and 3 respectively.

Find $E(W + X + Y)$, $Var(W + X + Y)$, $E(2W - X - Y)$ and $Var(2W - X - Y)$.

7 A piece of laminated plywood consists of three pieces of wood of type A and two pieces of type B. The thickness of A has mean 2 mm and variance 0.04 mm^2. The thickness of B has mean 1 mm and variance 0.01 mm^2. Find the mean and variance of the thickness of the laminated plywood.

8 The random variable S is the score when an ordinary fair dice is thrown. The random variable T is the number of tails obtained when a fair coin is tossed once.

Find $E(S)$, $Var(S)$, $E(6T)$, $Var(6T)$, $E(S + 6T)$ and $Var(S + 6T)$.

9 The random variable X has the probability distribution shown in the table.

x	1	2	3	4
$P(X = x)$	$\frac{3}{8}$	$\frac{1}{4}$	$\frac{1}{4}$	$\frac{1}{8}$

Find the mean and variance of the distribution of the sum of three independent observations of X.

10 In the game of American football, players are rated according to their performance. The quarterback's rating, Q, is calculated using the formula $Q = \frac{5}{6}(C + 5Y + 2.5 + 4T - 5I)$, where C, Y, T and I are variables which can be considered independent with means and variances as shown in the table.

Variable	C	Y	T	I
Mean	60.0	6.8	4.5	3.1
Variance	68.5	2.3	9.0	7.1

C, T and I are the percentage completions, touchdown passes and interceptions per pass attempt. Y is the number of metres gained divided by the passes attempted.

Find $E(Q)$ and $Var(Q)$.

11 The random variable Y which can only take the values 0 and 1 is called the Bernoulli distribution. Given that $P(1) = p$, show that $E(Y) = p$ and $Var(Y) = p(1 - p)$.

The binomial distribution can be considered to be a series of Bernoulli trials. That is,

$$X = Y_1 + Y_2 + \ldots + Y_n.$$

Show that $E(X) = np$ and $Var(X) = np(1 - p)$.

2.4 Linear functions and combinations of normally distributed random variables

In S1 Chapter 9 you saw that the normal distribution can be used to model the distribution of a continuous random variable. A normally distributed random variable has the useful and interesting property that a linear function of it is also normally distributed. In addition, linear combinations of independent normal variables are also normally distributed.

> If a continuous random variable X is distributed normally, then $aX + b$ (where a and b are constants) is also distributed normally.
>
> If continuous random variables X, Y, ... have independent normal distributions, then $aX + bY + \ldots$ (where a, b, ... are constants) is also distributed normally.

The following examples illustrate applications of these important properties.

Example 2.4.1

The mass of an empty lift cage is 210 kg. If the masses (in kg) of adults are distributed as $N(70, 950)$, what is the probability that the mass of the lift cage containing 10 adults chosen at random exceeds 1000 kg?

Let the mass in kg of an adult chosen at random be X. Then $X \sim N(70, 950)$.

Let the mass of the cage containing 10 adults be M. Then

$$M = 210 + X_1 + X_2 + \ldots + X_{10}.$$

Assuming that the masses of the adults are independent, M will be normally distributed with mean and variance given by

$$\begin{aligned}
\mathrm{E}(M) &= \mathrm{E}(210 + X_1 + X_2 + \ldots + X_{10}) \\
&= 210 + \mathrm{E}(X_1) + \mathrm{E}(X_2) + \ldots + \mathrm{E}(X_{10}) \\
&= 210 + 10 \times \mathrm{E}(X) \\
&= 210 + 10 \times 70 = 910,
\end{aligned}$$

and
$$\begin{aligned}
\mathrm{Var}(M) &= \mathrm{Var}(210 + X_1 + X_2 + \ldots + X_{10}) \\
&= \mathrm{Var}(X_1) + \mathrm{Var}(X_2) + \ldots + \mathrm{Var}(X_{10}) \\
&= 10 \times \mathrm{Var}(X) \\
&= 10 \times 950 = 9500.
\end{aligned}$$

So $M \sim N(910, 9500)$.

As the masses of the adults are independent $M \neq 210 + 10X$. *So* $\mathrm{Var}(M) \neq 10^2 \mathrm{Var}(X)$.

$$\begin{aligned}
\mathrm{P}(M > 1000) &= \mathrm{P}\left(Z > \frac{1000 - 910}{\sqrt{9500}}\right) = \mathrm{P}(Z > 0.923) \\
&= 1 - \Phi(0.923) \\
&= 1 - 0.8220 = 0.1780.
\end{aligned}$$

The probability that the mass of the lift cage with 10 adults exceeds 1000 kg is 0.178, correct to 3 significant figures.

Example 2.4.2

Jamal rents a phone under a scheme which has a fixed charge of $8 per month with calls charged at $0.20 per minute. Selina rents her phone under a different scheme. This has a fixed charge of $20 with calls charged at $0.10 per minute. The number of minutes that Jamal uses his phone in a randomly chosen month is denoted by J and the number of minutes that Selina uses her phone in a randomly chosen month is denoted by S. It is given that $J \sim N(120, 49)$ and $S \sim N(130, 25)$, and that J and S are independent.

(a) Find the distribution of the amount spent by Jamal on his phone in a randomly chosen month.

(b) Find the distribution of the amount spent by Selina on her phone in a randomly chosen month.

(c) Find the probability that in a randomly chosen month Jamal pays more for his phone than Selina.

(a) Let C denote the amount spent by Jamal on his phone in a randomly chosen month. Then $C = 8 + 0.2J$.

Since J is normally distributed, C is also normally distributed with

$$E(C) = E(8 + 0.2J) = 8 + 0.2E(J) = 8 + 0.2 \times 120 = 32$$

and $Var(C) = Var(8 + 0.2J) = 0.2^2 Var(J) = 0.04 \times 49 = 1.96.$

So $C \sim N(32, 1.96)$.

(b) Let D denote the amount spent by Selina on her phone in a randomly chosen month. Then $D = 20 + 0.1S$.

Since S is normally distributed, D is also normally distributed with

$$E(D) = E(20 + 0.1S) = 20 + 0.1E(S) = 20 + 0.1 \times 130 = 33$$

and $Var(D) = Var(20 + 0.1S) = 0.1^2 Var(S) = 0.01 \times 25 = 0.25.$

So $D \sim N(33, 0.25)$.

(c) In order for Jamal to pay more than Selina in a randomly chosen month, it is necessary that $C > D$. This inequality can also be expressed as $C - D > 0$, so the problem can be solved by considering the distribution of $C - D$. Since J and S are independent, C and D are also independent. Thus $C - D$ is normally distributed with

$$E(C) - E(D) = 32 - 33 = -1$$

and $Var(C - D) = Var(C) + Var(D) = 1.96 + 0.25 = 2.21.$

So $C - D \sim N(-1, 2.21)$.

Jamal pays more than Selina in a randomly chosen month if $C - D > 0$.

$$P(C - D > 0) = P\left(Z > \frac{0 - (-1)}{2.21} \right) = P(Z > 0.673)$$
$$= 1 - \Phi(0.673)$$
$$= 1 - 0.7946$$
$$= 0.2504.$$

The probability that Jamal pays more than Selina in a randomly chosen month is 0.250, correct to 3 significant figures.

2.5 The distribution of the sum of two independent Poisson variables

Suppose you have two Poisson variables, $X \sim \text{Po}(\lambda_X)$ and $Y \sim \text{Po}(\lambda_Y)$.

Then

$$E(X+Y) = E(X) + E(Y) = \lambda_X + \lambda_Y.$$

Since X and Y are Poisson variables, their variances are equal to their corresponding means. Thus, provided that X and Y are independent,

$$\text{Var}(X+Y) = \text{Var}(X) + \text{Var}(Y) = \lambda_X + \lambda_Y.$$

Thus the mean and variance of $X+Y$ are also equal. This means that it is *possible* for $X+Y$ to have a Poisson distribution. In fact, it can be *proved* that $X+Y$ has a Poisson distribution, provided that X and Y are independent, but a proof will not be given here.

Note that a linear combination of independent Poisson variables of the form $aX + bY$ (where a and b take values other than 1) cannot have a Poisson distribution since in that case the mean $(a\lambda_X + b\lambda_Y)$ is not equal to the variance $(a^2\lambda_X + b^2\lambda_Y)$.

Example 2.5.1

The numbers of emissions per minute from two radioactive sources are modelled by independent random variables X and Y which have Poisson distributions with means 5 and 8 respectively. Calculate the probability that in any minute the total number of emissions from the two sources is less than 6.

The total number of emissions, $X+Y$, has a Poisson distribution with mean $5 + 8 = 13$.

$$P(X < 6) = P(X=0) + P(X=1) + P(X=2) + P(X=3) + P(X=4) + P(X=5)$$

$$= e^{-13} + e^{-13} \times 13 + e^{-13} \times \frac{13^2}{2!} + e^{-13} \times \frac{13^3}{3!} + e^{-13} \times \frac{13^4}{4!} + e^{-13} \times \frac{13^5}{5!}$$

$$= 0.010\ 733\ldots.$$

The probability that in any minute the total number of emissions from the two sources is less than 6 is 0.0107, correct to 3 significant figures.

Since $X+Y$ has a normal distribution when X and Y have independent normal distributions and $X+Y$ has a Poisson distribution when X and Y have independent Poisson distributions you may be wondering whether these are particular instances of a general rule. This is not the case. If two independent random variables follow the same type of distribution, it is not necessarily true that their sum also follows this type of distribution.

For example, if X and Y have binomial distributions with different values of p, $X+Y$ does not have a binomial distribution; if X and Y have uniform distributions, $X+Y$ does not have a uniform distribution.

Exercise 2B

1 The heights of a population of male students are distributed normally with mean 178 cm and standard deviation 5 cm. The heights of a population of female students are distributed normally with mean 168 cm and standard deviation 4 cm. Find the probability that a randomly chosen female is taller than a randomly chosen male.

2 W is the mass of lemonade in a fully filled bottle, B is the mass of the bottle and C is the mass of the crate into which 12 filled bottles are placed for transportation, all in grams. It is given that $W \sim N(825,15^2)$, $B \sim N(400,10^2)$ and $C \sim N(1500,20^2)$. Find the probability that a fully filled crate weighs less than 16.1 kg.

3 The times of four athletes for the 400 m are each distributed normally with mean 47 seconds and standard deviation 2 seconds. The four athletes are to compete in a 4×400 m relay race. Find the probability that their total time is less than 3 minutes.

4 The capacities of small bottles of perfume are distributed normally with mean 50 ml and standard deviation 3 ml. The capacities of large bottles of the same perfume are distributed normally with mean 80 ml and standard deviation 5 ml. Find the probability that the total capacity of 3 small bottles is greater than the total capacity of 2 large bottles.

5 The diameters of a consignment of bolts are distributed normally with mean 1.05 cm and standard deviation 0.1 cm. The diameters of the holes in a consignment of nuts are distributed normally with mean 1.1 cm and standard deviation 0.1 cm. Find the probability that a randomly chosen bolt will not fit inside a randomly chosen nut.

6 The amount of black coffee dispensed by a drinks machine is distributed normally with mean 200 ml and standard deviation 5 ml. If a customer requires white coffee, milk is also dispensed. The amount of milk is distributed normally with mean 20 ml and standard deviation 2 ml. Find the probability that the total amount of liquid dispensed when a customer chooses white coffee is less than 210 ml.

7 Given that $X \sim N(\mu,10)$, $Y \sim N(12,\sigma^2)$ and $3X - 4Y \sim N(0,234)$, find μ and σ^2.

8 You are given that $X \sim Po(3)$ and $Y \sim Po(2)$. Find the mean and variance of
(a) $X + Y$, (b) $X - Y$, (c) $3X + 2$.
Which of (a), (b) and (c) has a Poisson distribution?

9 The number of vehicles travelling on a particular road towards a town centre has a Poisson distribution with mean 6 per minute. The number of vehicles travelling away from the town centre on the same road at the same time of day has a Poisson distribution with mean 3 per minute. Find the probability that the total number of vehicles seen passing a given point in a 1-minute period is less than 6.

10 The number of goals scored per match by a football team during a season has a Poisson distribution with mean 1.5. The number of goals conceded per match by the same team during the same season has a Poisson distribution with mean 1. Find the probability that a match involving the team produced more than 3 goals.

Miscellaneous exercise 2

1 The random variable X has mean 0 and variance $\frac{4}{5}$. $Y = aX + b$. It is given that $E(Y) = 50$ and $\text{Var}(Y) = 80$. Find a and b. (OCR)

2 The random variable X takes values $-2, 0, 2$ with probabilities $\frac{1}{4}, \frac{1}{2}, \frac{1}{4}$ respectively. Find $\text{Var}(X)$ and $E(|X|)$.

The random variable Y is defined by $Y = X_1 + X_2$, where X_1 and X_2 are two independent observations of X. Find $\text{Var}(Y)$ and $E(Y + 3)$. (OCR)

3 The probability of there being X unusable matches in a full box of Surelite matches is given by

$$P(X = 0) = 8k, \quad P(X = 1) = 5k, \quad P(X = 2) = P(X = 3) = k, \quad P(X \geq 4) = 0.$$

Determine the constant k and the expectation and variance of X. Two full boxes of Surelite matches are chosen at random and the total number Y of unusable matches is determined. State the values of the expectation and variance of Y. (OCR)

4 In a packaging factory, the empty containers for a certain product have a mean weight of 400 g with a standard deviation of 10 g. The mean weight of the contents of a full container is 800 g with a standard deviation of 15 g. Find the expected total weight of 10 full containers and the standard deviation of this weight, assuming that the weights of containers and contents are independent.

Assuming further that these weights are normally distributed random variables, find the proportion of batches of 10 full containers which weigh more than 12.1 kg.

If 1% of the containers are found to be holding weights of product which are less than the guaranteed minimum amount, deduce this minimum weight. (OCR)

5 During a weekday, heavy lorries pass a census point P on a village high street independently and at random times. The mean rate for westward travelling lorries is 2 in any 30-minute period, and for eastward travelling lorries is 3 in any 30-minute period.

Find the probability

(a) that there will be no lorries passing P in a given 10-minute period,

(b) that at least one lorry from each direction will pass P in a given 10-minute period,

(c) that there will be exactly 4 lorries passing P in a given 20-minute period. (OCR)

6 Telephone calls reach a secretary independently and at random, internal ones at a mean rate of 2 in any 5-minute period, and external ones at a mean rate of 1 in any 5-minute period. Calculate the probability that there will be more than 2 calls in any period of 2 minutes. (OCR)

7 The mass of tea in Supacuppa tea-bags has a normal distribution with mean 4.1 g and standard deviation 0.12 g. The mass of tea in Bumpacuppa tea-bags has a normal distribution with mean 5.2 g and standard deviation 0.15 g.

   (a) Find the probability that five randomly chosen Supacuppa tea-bags contain a total of more than 20.8 g of tea.

   (b) Find the probability that the total mass of tea in five randomly chosen Supacuppa tea-bags is more than the total mass of tea in four randomly chosen Bumpacuppa tea-bags.   (OCR)

8 The independent random variables R and S each have normal distributions. The means of R and S are 10 and 12 respectively, and the variances are 9 and 16 respectively. Find the following probabilities, giving your answer correct to 3 significant figures:

   (a) $P(R < S)$,

   (b) $P(2R > S_1 + S_2)$, where S_1 and S_2 are two independent observations of S.   (OCR)

9 Small packets of nails are advertised as having average weight 500 g, and large packets as having average weight 1000 g. Assume that the packet weights are distributed normally with means as advertised, and standard deviations of 10 g for a small packet and 15 g for a large packet. Giving your answers correct to 3 decimal places,

   (a) find the probability that two randomly chosen small packets have a total weight between 990 g and 1020 g,

   (b) find the probability that the weight of one randomly chosen large packet exceeds the total weight of two randomly chosen small packets by at least 25 g,

   (c) find the probability that one half of the weight of one randomly chosen large packet exceeds the weight of one randomly chosen small packet by at least 12.5 g.   (OCR)

10 The random variable X has a normal distribution with mean 3 and variance 4. The random variable S is the sum of 100 independent observations of X, and the random variable T is the sum of a further 300 independent observations of X. Giving your answers to 3 decimal places, find

   (a) $P(S > 310)$,   (b) $P(3S > 50 + T)$.

The random variable N is the sum of n independent observations of X. State the approximate value of $P(N > 3.5n)$ as n becomes very large, justifying your answer.

                                                                         (OCR)

3 Continuous random variables

This chapter looks at the way in which continuous random variables are modelled mathematically. When you have completed it you should

- understand what a continuous random variable is
- know the properties of a probability density function and be able to use them
- be able to use a probability density function to solve problems involving probabilities
- be able to find the median and other percentiles of a distribution in simple cases
- be able to calculate the mean and variance of a distribution.

3.1 Defining the probability density function of a continuous random variable

In S1 Section 9.2, you met the normal distribution as a model for a continuous variable. A normal distribution is illustrated in Fig. 3.1 Although many continuous variables can be modelled by this distribution, there are also many which cannot.

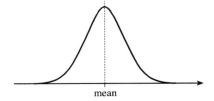

Fig. 3.1. A normal distribution.

Consider the following example. A ski-lift takes you from the bottom to the top of a ski-slope. The 'cars' on the ski-lift are attached at equal intervals along a cable which travels at a fixed speed. The speed of the cars is so low that people can step in and out of the cars at the station without the cars having to stop. The time interval between one car and the next arriving at the station is 5 minutes. You do not know the timetable for the cars and so you turn up at the station at a random time and wait for a car. Your waiting time, X (measured in minutes), is an example of a random variable because its value depends on chance. However, it is also a continuous variable because the waiting time can take any value in the interval 0 to 5 minutes, that is $0 \leqslant X < 5$.

As was explained in S1 Chapter 9, continuous variables are modelled by specifying the probability that the variable lies within a particular interval.

For example, you would expect that $P(0 \leqslant X < 2.5) = \frac{1}{2}$ since the interval from 0 to 2.5 accounts for half the values which X can take, and all values are equally likely. Extending this idea, you would expect $P(0 \leqslant X < 1) = \frac{1}{5}$ since this interval covers $\frac{1}{5}$ of the total interval. Similarly $P(1 \leqslant X < 2) = P(2 \leqslant X < 3) = P(3 \leqslant X < 4) = P(4 \leqslant X < 5) = 0.2$. These probabilities could be represented on a diagram similar to a histogram, as shown in Fig. 3.2. In this diagram the area of each block gives the probability that X lies in the corresponding interval. Since the width of each block is 1 its height must be 0.2 in order to make the area equal to 0.2. Notice that the total area under the curve must be 1 since the probabilities must sum to 1.

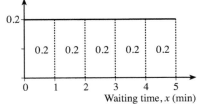

Fig. 3.2. Diagram to represent the probabilities of waiting times for a ski-lift.

The choice of the intervals 0 to 1, 1 to 2 and so on is arbitrary. In order to make the model more general you need to be able to find the probability that X lies within any given interval. Suppose that the divisions between the blocks in Fig. 3.2 are removed so as to give Fig. 3.3. In this diagram, probabilities still correspond to areas. For example, $P(0.5 < X \leqslant 1)$ is given by the area under the curve between 0.5 and 1. This is shown as the shaded area in Fig. 3.4: its value is 0.1 as you would expect.

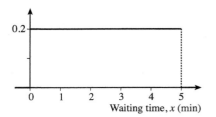

Fig. 3.3. Diagram to represent the distribution of waiting times for a ski-lift.

You should now be able to see that the probability distribution of the waiting times can be modelled by the continuous function $f(x)$ which describes Fig. 3.3. The required function is

$$f(x) = \begin{cases} 0.2 & \text{for } 0 \leqslant x < 5, \\ 0 & \text{otherwise.} \end{cases}$$

It is usual to define $f(x)$ *for all real values of* x.

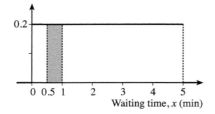

Fig. 3.4. Diagram to represent the probability of waiting between 0.5 and 1.0 minutes for the lift.

Once $f(x)$ has been defined, you can find probabilities by calculating areas below the curve $y = f(x)$. For example, $P(1.3 \leqslant X \leqslant 3.5) = 0.2 \times (3.5 - 1.3) = 0.44$.

Notice that, if you want the probability that $X = 1.3$ or the probability that $X = 3.5$, the answer is zero. These are just single instants of time, and although it is theoretically possible that a car may arrive at either of those instants, the probability is actually zero. This means that $P(1.3 < X < 3.5) = P(1.3 \leqslant X < 3.5) = P(1.3 < X \leqslant 3.5) = P(1.3 \leqslant X \leqslant 3.5)$. This situation is characteristic of continuous distributions.

The function $f(x)$ is called a **probability density function**. It cannot take negative values because probabilities are never negative. It must also have the property that the total area under the curve $y = f(x)$ is equal to one. This is because this area represents the probability that X takes any real value and this probability must be 1.

The suitability of this function as a model for actual waiting times could be tested by collecting some data and comparing a histogram of these experimental results with the shape of $y = f(x)$. Some results are given in Table 3.5.

Waiting time, x (min)	Frequency	Relative frequency	Class width	Relative frequency density
$0 \leqslant x < 1$	107	0.214	1	0.214
$1 \leqslant x < 2$	98	0.196	1	0.196
$2 \leqslant x < 3$	105	0.210	1	0.210
$3 \leqslant x < 5$	190	0.380	2	0.190

Table 3.5. Waiting times for the ski-lift for a sample of 500 people.

The third column gives the relative frequencies: these are found by dividing each frequency by the total frequency, in this case 500. The relative frequency gives the experimental probability that the waiting time lies in a given interval. The fifth column gives the relative frequency density: this is found by dividing the relative frequency by the class width. The data are illustrated by the histogram in Fig. 3.6.

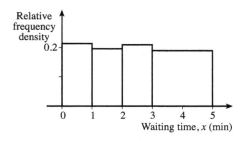

Fig. 3.6. Histogram of relative frequency for the data in Table 3.5.

Normally a histogram is plotted with frequency density rather than *relative* frequency density on the vertical axis. The reason for using relative frequency density in this case (and others in this section) is that area then represents *relative* frequency and hence experimental probability. As a result a direct comparison can be made between this diagram and Fig. 3.3, in which area represents theoretical probability. You can see that the diagrams are very similar. The experimental probabilities are not exactly equal to the theoretical ones. For example, the experimental probability of waiting between 1 and 2 minutes is 0.196 whereas theoretically it is 0.2. This is not surprising: you saw in S1 Section 4.1 that a probability model aims to describe what happens 'in the long run'.

The model for waiting times for a ski-lift was found by theoretical arguments and then confirmed experimentally. However, often it is not possible to predict the form of the probability density function. Instead, data are collected and the shape of the resulting histogram of the relative frequency density may suggest the form of the probability density function. Table 3.7 gives some more data; these relate to the time interval between one patient and the next going into the consulting room at a doctor's surgery. The doctor's receptionist always allows at least a 5-minute interval between the start of one consultation and the start of the next.

Time interval, x (min)	Frequency	Relative frequency	Class width	Relative frequency density
$5 \leqslant x < 6$	16	0.16	1	0.160
$6 \leqslant x < 7$	14	0.14	1	0.140
$7 \leqslant x < 8$	8	0.08	1	0.080
$8 \leqslant x < 9$	9	0.09	1	0.090
$9 \leqslant x < 10$	9	0.09	1	0.090
$10 \leqslant x < 11$	8	0.08	1	0.080
$11 \leqslant x < 13$	12	0.12	2	0.060
$13 \leqslant x < 15$	6	0.06	2	0.030
$15 \leqslant x < 20$	10	0.10	5	0.020
$20 \leqslant x < 25$	6	0.06	5	0.012

Table 3.7. Time intervals between patients entering the consulting room, for 100 patients.

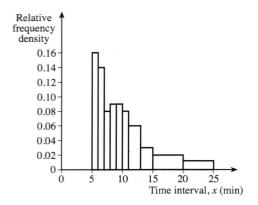

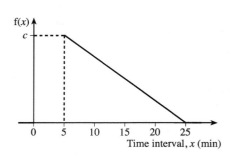

Fig. 3.8. Histogram of relative frequency for the data in Table 3.7.

Fig. 3.9. Simple model for the time intervals between patients entering the consulting room.

Fig. 3.8 shows a histogram of the relative frequency densities. The shape of this histogram suggests that a very simple model to describe this situation might be the straight line segment shown in Fig. 3.9. This line has been drawn to cut the horizontal axis at 25 since all the time intervals were less than this value. You can find the equation of this line by remembering that the area under the graph of $f(x)$ must be 1, since the total probability must be 1. Let $f(5)$ equal c. Then, since the region under the graph of $f(x)$ is a triangle of area 1,

$$\tfrac{1}{2} \times 20 \times c = 1, \quad \text{giving} \quad c = \tfrac{1}{10}.$$

Thus the gradient of the line is given by $\dfrac{0 - \tfrac{1}{10}}{25 - 5} = -\tfrac{1}{200}$ and the equation of the line is

$$y - \tfrac{1}{10} = -\tfrac{1}{200}(x - 5), \quad \text{or} \quad y = -\tfrac{1}{200}x + \tfrac{1}{8}.$$

The probability density function is therefore

$$f(x) = \begin{cases} -\tfrac{1}{200}x + \tfrac{1}{8} & \text{for } 5 \leq x \leq 25, \\ 0 & \text{otherwise.} \end{cases}$$

This model can then be used to find probabilities. For example the probability of a time interval of more than 17 minutes is given by the area of the shaded region in Fig. 3.10. This can be found by using simple geometry.

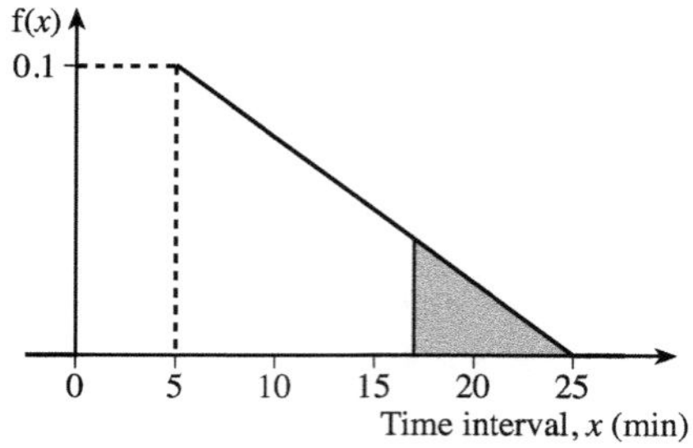

When $x = 17$, $f(x) = -\tfrac{1}{200} \times 17 + \tfrac{1}{8} = \tfrac{1}{25}$.

So $P(X > 17) = \tfrac{1}{2} \times 8 \times \tfrac{1}{25} = \tfrac{4}{25}$.

Fig. 3.10. Diagram to illustrate the theoretical probability of a time interval of more than 17 minutes.

Alternatively you can use integration to find the area as follows.

$$P(X > 17) = \int_{17}^{25} \left(-\tfrac{1}{200}x + \tfrac{1}{8}\right) dx = \left[-\tfrac{1}{400}x^2 + \tfrac{1}{8}x\right]_{17}^{25}$$

$$= \left(-\tfrac{1}{400} \times 25^2 + \tfrac{1}{8} \times 25\right) - \left(-\tfrac{1}{400} \times 17^2 + \tfrac{1}{8} \times 17\right)$$

$$P(X > 17) = \left(-\frac{625}{400} + \frac{25}{8}\right) - \left(-\frac{289}{400} + \frac{17}{8}\right)$$

$$= \left(-\frac{625}{400} + \frac{289}{400}\right) + \left(\frac{25}{8} - \frac{17}{8}\right) = -\frac{336}{400} + 1 = \frac{4}{25}, \text{ as before.}$$

You should be able to deduce the value of
$P(X \leqslant 17)$ *without further detailed calculation.*

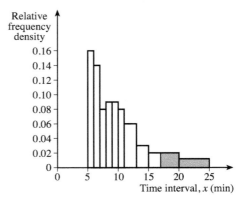

Fig. 3.11 reproduces Fig. 3.8. The experimental value of $P(X > 17)$ is given by the shaded area which is equal to

$$(3 \times 0.02) + (5 \times 0.012) = 0.12 = \frac{3}{25}.$$

This agrees quite well with the theoretical value.

Fig. 3.11. Diagram to illustrate the experimental probability of a time interval of more than 17 minutes.

Perhaps you can see a weakness in the model for time intervals in the doctor's surgery. The model is based on a small data set in which no value is greater than 25 minutes. As a result the model predicts that the time interval can *never* exceed 25 minutes. However, in the future it might be longer. Fig. 3.12 is a histogram of the relative frequency density for results collected from a larger sample of patients. You can see that quite a few values are greater than 25.

This histogram also shows another weakness in the original model: it looks as though a curve would fit the data better than a straight line. For example, a function of the type $f(x) = \dfrac{k}{x^n}$ where n and k are positive constants might be more suitable.

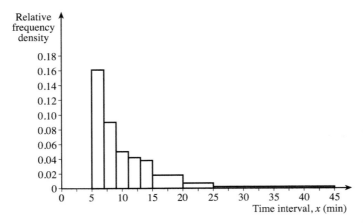

Fig. 3.12. Histogram of relative frequency for a larger sample of time intervals.

It turns out that the function $f(x) = \dfrac{5}{x^2}$ for $x \geqslant 5$ fits the histogram quite well.

Fig. 3.13 shows the histogram with the graph of this function superimposed on it. You can see that the curve and the histogram have similar shapes. This model has the advantage that it sets no upper limit to the time interval.

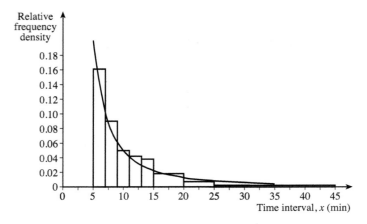

Fig. 3.13. Histogram in Fig. 3.12 with $f(x)$ superimposed.

Since $f(x)$ is always greater than zero it has one of the required properties of a probability density function. In order to accept this function as a probability density function, you also need to check that the area of the region underneath its graph is 1. This can be done by integration as follows.

$$\text{Area} = \int_{-\infty}^{\infty} f(x)\,dx = \int_{-\infty}^{5} 0\,dx + \int_{5}^{\infty} \frac{5}{x^2}\,dx$$

$$= 0 + \left[-\frac{5}{x} \right]_{5}^{\infty} = -0 - (-1) = 1.$$

So the function

$$f(x) = \begin{cases} \dfrac{5}{x^2} & \text{for } x \geqslant 5, \\ 0 & \text{otherwise,} \end{cases}$$

has the properties required of a probability density function.

Here is a summary of the properties of a probability density function.

The **probability density function**, $f(x)$, of a continuous random variable X is defined for all real values of x. It has the properties:

(a) $f(x) \geqslant 0$ for all x,   (b) $\displaystyle\int_{-\infty}^{\infty} f(x)\,dx = 1.$

The probability that X lies in the interval $a \leqslant x \leqslant b$ is given by the area under the graph of $f(x)$ between a and b. This area can sometimes be found by using geometrical properties or it can be found from the integral

$$P(a \leqslant X \leqslant b) = \int_{a}^{b} f(x)\,dx.$$

Example 3.1.1

The continuous random variable X has the probability density function given by

$$f(x) = \begin{cases} k(1+x^2) & \text{for } -1 \leqslant x \leqslant 1, \\ 0 & \text{otherwise,} \end{cases}$$

where k is a constant.

(a) Find the value of k.   (b) Find $P(0.3 \leqslant X \leqslant 0.6)$.   (c) Find $P(|X| < 0.2)$.

(a) $\displaystyle\int_{-\infty}^{\infty} f(x)\,dx = \int_{-1}^{1} k(1+x^2)\,dx = k\left[x + \tfrac{1}{3}x^3\right]_{-1}^{1}$

$\qquad\qquad = k\left(1 + \tfrac{1}{3} \times 1^3\right) - k\left((-1) + \tfrac{1}{3} \times (-1)^3\right) = k \times \tfrac{4}{3} - k \times \left(-\tfrac{4}{3}\right) = \tfrac{8}{3}k.$

Using the second property in the box,

$$\int_{-\infty}^{\infty} f(x)\,dx = 1, \text{ so } \tfrac{8}{3}k = 1. \text{ This gives } k = \tfrac{3}{8}.$$

(b) $P(0.3 \leqslant X \leqslant 0.6) = \displaystyle\int_{0.3}^{0.6} \tfrac{3}{8}(1+x^2)\,dx = \tfrac{3}{8}\left[x + \tfrac{1}{3}x^3\right]_{0.3}^{0.6}$

$\qquad\qquad = \tfrac{3}{8}\left(0.6 + \tfrac{1}{3} \times 0.6^3\right) - \tfrac{3}{8}\left(0.3 + \tfrac{1}{3} \times 0.3^3\right)$

$\qquad\qquad = 0.136, \text{ correct to 3 significant figures.}$

(c) $P(|X| < 0.2) = P(-0.2 < X < 0.2) = \displaystyle\int_{-0.2}^{0.2} \tfrac{3}{8}(1+x^2)\,dx = \tfrac{3}{8}\left[x + \tfrac{1}{3}x^3\right]_{-0.2}^{0.2}$

$\qquad\qquad = \tfrac{3}{8}\left(0.2 + \tfrac{1}{3} \times 0.2^3\right) - \tfrac{3}{8}\left((-0.2) + \tfrac{1}{3} \times (-0.2)^3\right) = 0.152.$

Example 3.1.2

It is proposed to model the annual salary, X, measured in thousands of $, paid to sales persons in a large company by the probability density function

$$f(x) = \begin{cases} cx^{-\frac{7}{2}} & \text{for } x \geqslant 16, \\ 0 & \text{otherwise.} \end{cases}$$

(a) Find the value of c.

(b) Find the probability that a person in this profession chosen at random earns between $20 000 and $30 000 per year.

(a) $\displaystyle\int_{-\infty}^{\infty} f(x)\,dx = \int_{16}^{\infty} cx^{-\frac{7}{2}}\,dx = \left[-\tfrac{2}{5}cx^{-\frac{5}{2}}\right]_{16}^{\infty} = (-0) - \left(-\tfrac{2}{5}c \times 16^{-\frac{5}{2}}\right) = \tfrac{1}{2560}c.$

Since $\displaystyle\int_{-\infty}^{\infty} f(x)\,dx = 1$, $\tfrac{1}{2560}c = 1$,   giving $c = 2560$.

(b) $P(20 \leqslant X \leqslant 30) = \int_{20}^{30} 2560 x^{-\frac{7}{2}} \, dx = \left[2560 \times \left(-\frac{2}{5} \right) x^{-\frac{5}{2}} \right]_{20}^{30}$

$\qquad = \left(2560 \times \left(-\frac{2}{5} \right) \times 30^{-\frac{5}{2}} \right) - \left(2560 \times \left(-\frac{2}{5} \right) \times 20^{-\frac{5}{2}} \right)$

$\qquad = 0.365$, correct to 3 significant figures.

Exercise 3A

In all the questions in this exercise, c and k are constants. In this and the following exercises, some questions involve the exponential function e^x. If you have not already met this function in P2 Chapter 4, you should omit these questions.

1 The probability density function $f(x) = \begin{cases} c\left(1 - \frac{1}{8}x\right) & 0 \leqslant x \leqslant 8, \\ 0 & \text{otherwise.} \end{cases}$

   (a) Find the value of the constant c.   (b) Find $P(X \geqslant 6)$.

   (c) Find $P(4 \leqslant X \leqslant 6)$.

2 The probability density function $f(x) = \begin{cases} kx^2 & 0 \leqslant x \leqslant 3, \\ 0 & \text{otherwise.} \end{cases}$

   (a) Find the value of the constant k.

   (b) Find $P(X \leqslant 2)$.

   (c) Find $P(1.5 \leqslant X \leqslant 2.5)$.

   (d) Given that the probability that X is less than h is 0.2, find the value of h, correct to 2 decimal places.

3 The probability density function $f(x) = \begin{cases} c\left(x^2 + 2\right) & 0 \leqslant x \leqslant 3, \\ 0 & \text{otherwise.} \end{cases}$

   (a) Find the value of the constant c.   (b) Find $P(X \leqslant 1.5)$.

4 The probability density function $f(x) = \begin{cases} c\left(4 - x^2\right) & -2 \leqslant x \leqslant 2, \\ 0 & \text{otherwise.} \end{cases}$

   (a) Find the value of the constant c.   (b) Find $P(X \geqslant 0)$.

   (c) Find $P(X \geqslant 1)$.   (d) Find $P(|X| \geqslant 1)$.

   (e) Find $P(-0.5 \leqslant X \leqslant 0.5)$.

5 The life, X, of the StayBrite light bulb is modelled by the probability density function

   $f(x) = \begin{cases} ke^{-2x} & x \geqslant 0, \\ 0 & \text{otherwise,} \end{cases}$

   where X is measured in thousands of hours.

   (a) Find k.

   (b) Find the probability that a StayBrite bulb lasts longer than 1000 hours.

   (c) Find the probability that a StayBrite bulb lasts less than 500 hours.

6 A printer ink cartridge has a life of X hours. The variable X is modelled by the
probability density function $f(x) = \begin{cases} kx^{-2} & x \geqslant 400, \\ 0 & \text{otherwise.} \end{cases}$

    (a) Find k.

    (b) Find the probability that such a cartridge has a life of at least 500 hours.

    (c) Find the probability that a cartridge will have to be replaced before 600 hours of use.

    (d) Find the probability that two cartridges will have to be replaced before each has been used for 600 hours.

7 The probability density function

$$f(x) = \begin{cases} k(x-a)^2 & 0 \leqslant x \leqslant a, \\ 0 & \text{otherwise,} \end{cases}$$

is shown in the sketch.

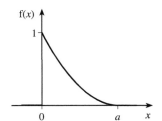

    (a) Use the information given in the sketch and the properties of probability density functions to find the values of a and k.

    (b) Find $P\left(X \geqslant \tfrac{1}{2}a\right)$.

3.2 The median of a continuous random variable

The median, M, of a continuous random variable is defined as the value which divides the area under the probability density function into two equal halves. Then the probability that X is above the median is equal to the probability that it is below. In mathematical terms the median is defined as follows.

> The median, M, of a continuous random variable is that value for which
> $$P(X \leqslant M) = \int_{-\infty}^{M} f(x)\,dx = \tfrac{1}{2}.$$

In simple cases the median can be found by considering symmetry. Fig. 3.14 reproduces Fig. 3.3 which showed the probability density function for the waiting times for the ski-lift. The line $x = 2.5$, which is shown on the diagram, divides the area under $f(x)$ in half and so the median waiting time, M, is 2.5 minutes.

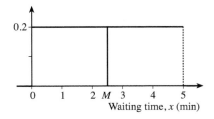

Fig. 3.14. Fig. 3.3 with the median shown.

Example 3.2.1

Two models are proposed for a garage's weekly sales, X, of petrol measured in units of 100 000 litres.

The first is $\qquad f(x) = \begin{cases} 2x & \text{for } 0 \leqslant x \leqslant 1, \\ 0 & \text{otherwise.} \end{cases}$

The second is $\qquad g(x) = \begin{cases} 12x^3(1-x^2) & \text{for } 0 \leqslant x \leqslant 1, \\ 0 & \text{otherwise.} \end{cases}$

(a) Find the median for the first model.

(b) Verify that the median of the second model is the same as that of the first model.

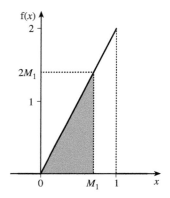

Fig. 3.15. Graph for part (a) of Example 3.2.1.　　　　Fig. 3.16. Graph for part (b) of Example 3.2.1.

(a) Fig. 3.15 shows the graph of $f(x)$. The median is denoted by M_1. The area of the shaded triangle is 0.5. When $x = M_1$, $f(x) = 2M_1$ so

$$\text{the area of shaded triangle} = \tfrac{1}{2} \times M_1 \times 2M_1 = 0.5,$$

giving $M_1 = \sqrt{\tfrac{1}{2}} = 0.707$, correct to 3 significant figures.

(b) Fig. 3.16 shows the graph of $g(x)$, with the median, M_1, for the first model marked. If the shaded area in Fig. 3.16 is equal to 0.5 then M_1 is also the median for the second model.

The area of the shaded region is

$$\int_0^{M_1} 12x^3(1-x^2)\,dx = \int_0^{M_1} (12x^3 - 12x^5)\,dx = \left[3x^4 - 2x^6\right]_0^{M_1}$$
$$= 3M_1{}^4 - 2M_1{}^6 = M_1{}^2\left(3M_1{}^2 - 2M_1{}^4\right).$$

Recalling that $M_1{}^2 = \tfrac{1}{2}$, the shaded area is $\tfrac{1}{2}\left(3 \times \tfrac{1}{2} - 2 \times \left(\tfrac{1}{2}\right)^2\right) = \tfrac{3}{4} - \tfrac{1}{4} = \tfrac{1}{2}$.

Thus the median of the second model is the same as the median of the first model.

Example 3.2.2
Find the median salary (to the nearest \$100) of the probability density function in Example 3.1.2.

Fig. 3.17 shows the graph of $f(x) = \begin{cases} 2560x^{-\frac{7}{2}} & \text{for } x \geqslant 16, \\ 0 & \text{otherwise.} \end{cases}$

The median is indicated by M and the shaded area is equal to 0.5.

Set the upper limit of the integral to M, form an equation and solve it for M as follows.

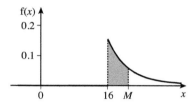

Fig. 3.17. Graph of the probability density function for Example 3.2.2.

$$P(X \leqslant M) = \int_{-\infty}^{M} 2560x^{-\frac{7}{2}} \, dx = \int_{16}^{M} 2560x^{-\frac{7}{2}} \, dx$$

$$= \left[2560 \times \left(-\tfrac{2}{5} \right) x^{-\frac{5}{2}} \right]_{16}^{M}$$

$$= \left(2560 \times \left(-\tfrac{2}{5} \right) \times M^{-\frac{5}{2}} \right) - \left(2560 \times \left(-\tfrac{2}{5} \right) \times 16^{-\frac{5}{2}} \right)$$

$$= -1024M^{-\frac{5}{2}} + 1.$$

This probability must equal 0.5, so $-1024M^{-\frac{5}{2}} + 1 = \tfrac{1}{2}$, giving $1024M^{-\frac{5}{2}} = \tfrac{1}{2}$.

Hence $M^{-\frac{5}{2}} = \tfrac{1}{2048}$, so $M^{\frac{5}{2}} = 2048$, giving $M = 21.1$, correct to 3 significant figures.

So the median salary is \$21 100 to the nearest \$100.

The method described above for finding the median can be extended to finding other percentiles. Suppose, for example, you wished to find the lower quartile, Q_1, of the distribution in Example 3.2.2. In this case

$$P(X \leqslant Q_1) = \int_{-\infty}^{Q_1} 2560x^{-\frac{7}{2}} \, dx = -1024Q_1^{-\frac{5}{2}} + 1 = \tfrac{1}{4}.$$

So $1024Q_1^{-\frac{5}{2}} = \tfrac{3}{4}$,   giving $Q_1 = \left(\dfrac{4096}{3} \right)^{\frac{2}{5}} = 18.0$, correct to 3 significant figures.

You can check that the upper quartile is 27.9, correct to 3 significant figures.

Exercise 3B

1 The probability density function $f(x) = \begin{cases} \tfrac{1}{9}x^2 & 0 \leqslant x \leqslant 3, \\ 0 & \text{otherwise.} \end{cases}$

Find the median value of X.

2 A printer ink cartridge has a life of X hours. The variable X is modelled by the probability density function $f(x) = \begin{cases} 400x^{-2} & x \geqslant 400, \\ 0 & \text{otherwise.} \end{cases}$

(a) Find the median lifetime of these cartridges.

(b) Find the value of b such that $P(X \leqslant b) = 0.6$.

3 The probability density function $f(x) = \begin{cases} 2x - 4 & 2 \leqslant x \leqslant 3, \\ 0 & \text{otherwise.} \end{cases}$

(a) Sketch the graph of $f(x)$.

(b) Find the median value of X.

(c) Find the interquartile range of X.

4 The life, X, of the StayBrite light bulb is modelled by the probability density function
$$f(x) = \begin{cases} 2e^{-2x} & x \geqslant 0, \\ 0 & \text{otherwise,} \end{cases}$$

where X is measured in thousands of hours.

(a) Sketch the graph of $f(x)$.

(b) Find the median life of these StayBrite bulbs.

(c) Find the value of w such that $P(X > w) = 0.9$.

5 The probability density function $f(x) = \begin{cases} \frac{2}{5}\left(1 - \frac{1}{5}x\right) & 0 \leqslant x \leqslant 5, \\ 0 & \text{otherwise.} \end{cases}$

(a) Sketch the graph of $f(x)$.   (b) Find the median value of X.

3.3 The expectation of a continuous random variable

The expectation (or mean) of a continuous random variable, X, is defined by

$$E(X) = \mu = \int_{-\infty}^{\infty} x f(x) \, dx. \qquad (3.1)$$

It is not possible to deduce this formula, but the following argument may help you to see why this definition makes sense.

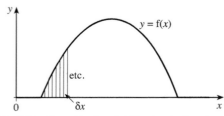

Fig. 3.18a. Generalised probability density function.

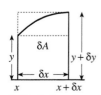

Fig. 3.18b. Enlarged version of one of the strips in Fig. 3.18a.

Look at Fig. 3.18a. This shows a probability density function, $f(x)$, for a continuous random variable, X. The region underneath $y = f(x)$ has been divided into narrow strips of width δx. Fig. 3.18b shows one of these strips, with some dotted lines added.

The probability that X takes a value between x and $x + \delta x$ is δA, the area of the strip.

Comparing with the equation for a discrete random variable $E(X) = \sum x P(X = x)$, this narrow strip makes a contribution to $E(X)$ which can be denoted by δE where

δE is between $x \delta A$ and $(x + \delta x) \times \delta A$.

Using the ideas which you met in P1 Section 16.2, you can say that δA lies between $y \delta x$ and $(y + \delta y) \delta x$,

so δE lies between $xy \delta x$ and $(x + \delta x)(y + \delta y) \delta x$.

Dividing through by δx gives

$\dfrac{\delta E}{\delta x}$ lies between xy and $(x + \delta x)(y + \delta y)$.

When δx tends to 0, $\dfrac{\delta E}{\delta x}$ tends to $\dfrac{dE}{dx}$. Also δy tends to 0, so that $y + \delta y$ tends to y. It follows that $\dfrac{dE}{dx} = xy$.

Since $y = f(x)$ this can also be written $\dfrac{dE}{dx} = x f(x)$.

Integrating,

$$E = \int_{-\infty}^{\infty} x f(x)\, dx.$$

Note the correspondence between this and the equation $E(X) = \sum x P(X = x)$ that for a discrete variable: $f(x)\, dx$ replaces $P(X = x)$ and the summation is replaced by an integral.

Example 3.3.1
Find the mean salary (to the nearest \$100) of the probability density function defined in Example 3.1.2 and compare it with the median salary which was calculated in Example 3.2.2.

From Equation 3.1,

$$\mu = E(X) = \int_{-\infty}^{\infty} x f(x)\, dx = \int_{16}^{\infty} x \times 2560 x^{-\frac{7}{2}}\, dx$$

$$= \int_{16}^{\infty} 2560 x^{-\frac{5}{2}}\, dx = \left[2560 \times \left(-\tfrac{2}{3}\right) x^{-\frac{3}{2}} \right]_{16}^{\infty}$$

$$= (0) - \left(2560 \times \left(-\tfrac{2}{3}\right) \times 16^{-\frac{3}{2}} \right) = 26 \tfrac{2}{3}.$$

To the nearest \$100, the mean salary is \$26 700.

The mean is greater than the median (= \$21 100) because the distribution is positively skewed.

3.4 The variance of a continuous random variable

> The variance of a continuous random variable, X, is given by
>
> $$\mathrm{Var}(X) = \sigma^2 = \int_{-\infty}^{\infty} x^2 f(x)\,dx - \mu^2. \qquad (3.2)$$

This formula shows the same parallel with the corresponding formula for a discrete random variable that was noted in the previous section. Recall (from S1 Section 8.2) that the variance of a discrete random variable is given by

$$\mathrm{Var}(X) = \sum x^2 P(X = x) - \mu^2.$$

As in the expectation formula $f(x)\,dx$ replaces $P(X = x)$ and an integral replaces the summation.

Example 3.4.1

For the continuous random variable X with probability density function defined by

$$f(x) = \begin{cases} \frac{3}{4}x(2-x) & \text{for } 0 \leqslant x \leqslant 2, \\ 0 & \text{otherwise,} \end{cases}$$

find   (a) the mean,   (b) the variance,   (c) $P(\mu - \sigma < X < \mu + \sigma)$.

(a) Using Equation 3.1

$$\mu = E(X) = \int_{-\infty}^{\infty} x f(x)\,dx = \int_{0}^{2} \frac{3}{4} x^2 (2-x)\,dx$$

$$= \int_{0}^{2} \left(\frac{3}{2} x^2 - \frac{3}{4} x^3 \right) dx = \left[\frac{3}{2} \times \frac{1}{3} x^3 - \frac{3}{4} \times \frac{1}{4} x^4 \right]_{0}^{2} = \left(\frac{1}{2} \times 2^3 - \frac{3}{16} \times 2^4 \right) = 4 - 3 = 1.$$

If you look at Fig. 3.19, which shows the graph of $y = f(x)$, you will see that there is a much quicker way of arriving at this result. Since this graph is symmetric about the line $x = 1$, the mean must be 1.

For a symmetrical probability density function, the mean is most easily found by using symmetry.

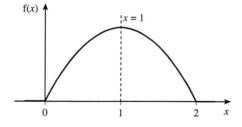

Fig. 3.19. Graph of the probability density function for Example 3.4.1.

(b) Using Equation 3.2

$$\sigma^2 = \mathrm{Var}(X) = \int_{-\infty}^{\infty} x^2 f(x)\,dx - \mu^2$$

$$= \int_{0}^{2} x^2 \times \frac{3}{4} x(2-x)\,dx - 1^2 = \int_{0}^{2} \left(\frac{3}{2} x^3 - \frac{3}{4} x^4 \right) dx - 1$$

$$= \left[\frac{3}{2} \times \frac{1}{4} x^4 - \frac{3}{4} \times \frac{1}{5} x^5 \right]_{0}^{2} - 1 = \left(6 - \frac{24}{5} \right) - (0) - 1 = \frac{1}{5} = 0.2.$$

(c) $P(\mu - \sigma < X < \mu + \sigma) = P(1 - \sqrt{0.2} < X < 1 + \sqrt{0.2})$

$$= \int_{1-\sqrt{0.2}}^{1+\sqrt{0.2}} \tfrac{3}{4} x(2-x)\, dx = \int_{1-\sqrt{0.2}}^{1+\sqrt{0.2}} \left(\tfrac{3}{2} x - \tfrac{3}{4} x^2 \right) dx$$

$$= \left[\tfrac{3}{2} \times \tfrac{1}{2} x^2 - \tfrac{3}{4} \times \tfrac{1}{3} x^3 \right]_{1-\sqrt{0.2}}^{1+\sqrt{0.2}}$$

$$= \left(\tfrac{3}{4}\left(1 + \sqrt{0.2}\right)^2 - \tfrac{1}{4}\left(1 + \sqrt{0.2}\right)^3 \right)$$

$$\quad - \left(\tfrac{3}{4}\left(1 - \sqrt{0.2}\right)^2 - \tfrac{1}{4}\left(1 - \sqrt{0.2}\right)^3 \right)$$

$$= \left(\tfrac{3}{4} \times 1.447...^2 - \tfrac{1}{4} \times 1.447...^3 \right)$$

$$\quad - \left(\tfrac{3}{4} \times 0.552...^2 - \tfrac{1}{4} \times 0.552...^3 \right)$$

$$= 0.626, \text{ correct to 3 significant figures.}$$

Exercise 3C

1 The probability density function $f(x) = \begin{cases} \tfrac{1}{9} x^2 & 0 \leqslant x \leqslant 3, \\ 0 & \text{otherwise.} \end{cases}$

Find the mean and variance of X.

2 The probability density function $f(x) = \begin{cases} 2x - 4 & 2 \leqslant x \leqslant 3, \\ 0 & \text{otherwise.} \end{cases}$

Find the mean and variance of X.

3 The probability density function $f(x) = \begin{cases} \tfrac{1}{4}\left(1 - \tfrac{1}{8} x\right) & 0 \leqslant x \leqslant 8, \\ 0 & \text{otherwise.} \end{cases}$

(a) Sketch the graph of $f(x)$.   (b) Find the mean and variance of X.

4 The mass, X kg, of silicon produced in a manufacturing process is modelled by the probability density function $f(x) = \begin{cases} \tfrac{3}{32}\left(4x - x^2\right) & 0 \leqslant x \leqslant 4, \\ 0 & \text{otherwise.} \end{cases}$

(a) Sketch the graph of $f(x)$.

(b) Find the mean and variance of the mass of silicon produced.

5 The EverOn battery has a life of X hours. The variable X is modelled by the probability density function $f(x) = \begin{cases} 3000x^{-4} & x \geqslant 10, \\ 0 & \text{otherwise.} \end{cases}$

(a) Sketch the graph of $f(x)$.

(b) Find the mean and variance of the lives of these EverOn torch batteries.

6 A printer ink cartridge has a life of X hours. The variable X is modelled by the probability density function $f(x) = \begin{cases} kx^{-2} & 400 \leqslant x \leqslant 900, \\ 0 & \text{otherwise.} \end{cases}$

(a) Sketch the graph of $f(x)$.   (b) Show that $k = 720$.

(c) Find the mean and variance of the lives of these cartridges.

Questions 7 and 8 require integration techniques covered in P3.

7 The life, X, of the StayBrite light bulb is modelled by the probability density function

$$f(x) = \begin{cases} 2e^{-2x} & x \geq 0, \\ 0 & \text{otherwise,} \end{cases}$$

where X is measured in thousands of hours.

Find the mean and variance of the lives of these StayBrite bulbs.

8 The radioactivity of krypton decays according to the probability model

$$f(x) = \begin{cases} ke^{-\lambda x} & x \geq 0, \\ 0 & \text{otherwise.} \end{cases}$$

(a) Show that $\lambda = k$.

(b) Find the mean and variance of X in terms of k.

Miscellaneous exercise 3

Questions 10 to 12 require integration techniques covered in P3.

1 A continuous random variable, X, has probability density function given by

$$f(x) = \begin{cases} \frac{1}{8}x & 0 \leq x \leq 4, \\ 0 & \text{elsewhere.} \end{cases}$$

(a) Calculate $P(X < 2)$.   (b) Calculate the expected value of X.   (OCR)

2 A continuous random variable, X, has the probability density function

$$f(x) = \begin{cases} \frac{1}{5} & 0 \leq x \leq 5, \\ 0 & \text{elsewhere.} \end{cases}$$

Find

(a) the mean, $E(X)$,

(b) $\text{Var}(X)$.   (OCR)

3 The time, in minutes, between two consecutive calls to a telephone switchboard is modelled by a continuous random variable, X. The probability density function, $f(x)$, for this random variable is given by

$$f(x) = \begin{cases} k(10 - x) & 0 \leq x \leq 10, \\ 0 & \text{otherwise.} \end{cases}$$

(a) Calculate the value of k.

(b) Find the mean time, $E(X)$, between two consecutive calls.

(c) Find $\text{Var}(X)$.   (OCR)

4 A continuous random variable, X, has the probability density function, $f(x)$, given by

$$f(x) = \begin{cases} k(4-x) & 0 \leqslant x \leqslant 4, \\ 0 & \text{otherwise.} \end{cases}$$

(a) Sketch the probability density function.

(b) Determine the value of k.

(c) Calculate the probability that $X > 2.5$.

(d) Find $E(X)$.   (OCR)

5 A continuous random variable, X, has the probability density function

$$f(x) = \begin{cases} \frac{1}{2}x & 0 \leqslant x \leqslant 2, \\ 0 & \text{otherwise.} \end{cases}$$

(a) Find the median value of X.

(b) Find $E(X)$.

(c) Find $\mathrm{Var}(X)$.   (OCR, adapted)

6 A continuous random variable, U, is uniformly distributed on $0.5 \leqslant u \leqslant 2.5$, as shown in the diagram.

(a) Find the probability density function $f(u)$.

(b) State the mean of U.

(c) Use integration to calculate the variance of U.

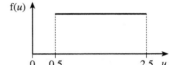

                                                                           (OCR)

7 The length, in metres, of 'offcuts' of wood found in a timber yard can be modelled by a continuous uniform distribution with density function, $f(x)$, defined as

$$f(x) = \begin{cases} \dfrac{1}{k} & 0.2 \leqslant x \leqslant 0.8, \\ 0 & \text{otherwise.} \end{cases}$$

(a) Write down the value of k.

(b) State the mean length.

(c) Calculate the variance of the length.

8 The random variable, X, has probability density function

$$f(x) = \begin{cases} kx^3 & 0 \leqslant x \leqslant 2, \\ 0 & \text{otherwise.} \end{cases}$$

(a) Find the value of k.

(b) Find $E(X)$.

(c) Find $\mathrm{Var}(X)$.

(d) Find the median of the distribution.

(e) Find the probability that an observation lies within one standard deviation of the mean.

                                                                           (OCR)

9 The random variable, X, has probability density function

$$f(x) = \begin{cases} \lambda x^3 & 0 \leqslant x \leqslant 4, \\ 0 & \text{otherwise.} \end{cases}$$

(a) Find the value of λ.

(b) Find $E(X)$.

(c) Find $\text{Var}(X)$.

(d) Find the probability $P(1 < X < 2)$. (OCR)

10 The continuous random variable, X, has probability density function

$$f(x) = \begin{cases} \dfrac{k}{x} & 1 \leqslant x \leqslant 2, \\ 0 & \text{otherwise.} \end{cases}$$

(a) Find the value of the constant k.

(b) Find the mean, $E(X)$.

(c) Find the variance, $\text{Var}(X)$.

(d) Determine the median value of X.

(e) Show that the probability that X is less than the mean is $-\dfrac{\ln(\ln 2)}{\ln 2}$. (OCR)

11 An internet surfer suggests that the time (t minutes) that he spends on the internet can be modelled by the probability density function

$$f(t) = \begin{cases} 0.1e^{-0.1t} & t \geqslant 0, \\ 0 & \text{otherwise.} \end{cases}$$

(a) Verify that this is a properly defined probability density function.

(b) Find the probability that the surfer spends less than 4 minutes on the internet.

(c) Find the probability that the surfer spends more than 10 minutes on the internet.

12 The random variable X has probability density function

$$f(x) = \begin{cases} ae^{-ax} & 0 \leqslant x, \\ 0 & \text{otherwise.} \end{cases}$$

(a) Find the median value of X.

The above distribution, with $a = 0.8$, is proposed as a model for the length of life, in years, of a species of bird.

(b) Find the expected number out of a total of 50 birds that would fall in the class interval 2–3 years.

13 The continuous random variable X has probability density function

$$f(x) = \begin{cases} (x - a)(2a - x) & a \leqslant x \leqslant 2a, \\ 0 & \text{otherwise.} \end{cases}$$

(a) Show that $a^3 = 6$.

(b) Find $E(X)$. (OCR)

14 A farmer needs to install a new water-pump. Pumps almost always run perfectly for the first year but thereafter if they fail they are not worth repairing and have to be replaced. They virtually never last more than 9 years. The length of time, in years, that the pumps last can be modelled by the continuous random variable X which has probability density function given by

$$f(x) = \begin{cases} \dfrac{k}{x} & 1 \leqslant x \leqslant 9, \\ 0 & \text{otherwise,} \end{cases}$$

where k is a constant.

(a) Show that $k = \dfrac{1}{2 \ln 3}$.

(b) Find the median length of life of a pump.

(c) Find the probability that a pump lasts between 1 and 2 years only.

(d) The farmer is offered a guarantee to cover the cost of replacing a pump that fails during the second year, at a cost of £300. Given that the pump will cost £1000 to replace if it fails during this year, what advice would you give the farmer about the merits of purchasing the guarantee?

(e) Pumps can be rented for an installation charge of £200 plus £250 per year, payable in advance. The yearly payment is not refundable if the pump fails before the end of the year. The farmer does not purchase the guarantee. Find the probability that a pump, at the end of its life, would have cost more to rent than to buy for £1000.

(OCR, adapted)

4 Sampling

This chapter looks at the ways of selecting a sample from a population. When you have completed it you should

- understand the distinction between a sample and a population
- understand how to select a random sample from a population
- appreciate the benefits of choosing a random sample
- recognise that a sample mean can be a random variable and use the facts that $E(\overline{X}) = \mu$ and that $\text{Var}(\overline{X}) = \dfrac{\sigma^2}{n}$
- use the fact that $\overline{X}$ has a normal distribution if X has a normal distribution
- understand the meaning of the central limit theorem and be able to use it in calculations.

4.1 Populations and samples

In statistics you usually wish to study a certain collection of individuals or items. The collection is known as the **population**. The term 'population' makes you think that the objects under consideration are living but this need not be true. For example, in S1 Section 1.1 you studied data taken from the internet about breakfast cereals. Breakfast cereals were the population in this case. It would have been useful to have information about all breakfast cereals produced from anywhere in the world, but this is unrealistic since it would take too long to do such an investigation and it would cost too much to carry out such a large survey. A more realistic aim would be to take a subsection of the population. The term used to describe a subsection of a population is a **sample**.

Ideally a sample should have all the characteristics of the larger population. In other words, it should be a miniature version of the population with identical properties. Unfortunately it is almost impossible for any sample of any population to possess all the characteristics of the population. For example, in the investigation into breakfast cereals the information was collected in the US, so many breakfast cereals produced in other countries were not represented. It is quite likely that other cereals have different properties from those produced and sold in the US but no information was available for them. Occasionally you may have the time, money and resources to investigate a population completely. Such an investigation is called a **census**. The governments of many countries carry out censuses of their populations. Census data can be used in planning, for example, the provision of housing and education.

In general the selection of a sample is preferable to a census, because it takes too much time and costs too much money to do a census. There are several other reasons why you might prefer to take a sample rather than carry out a census.

- There are likely to be fewer recording errors when a sample is taken because it will probably involve fewer people. Those people are likely to have more time to carry out the measurement and recording for the survey because the size and complexity of the task will not be as great as with a census.

- The analysis of the results can be carried out more speedily.
- Sometimes measurement on an individual sample member may result in its destruction. For example, if you were recording the lifetimes of batteries of a certain brand, then every battery which was selected would be used until it no longer worked. If a census were done then every battery ever made by that manufacturer would be destroyed. It is unlikely that any manufacturer would be willing to agree to such a census!

When you take a sample from a population you want to try to ensure that your method of selection is done in such a way that no feature of the population is over-represented or under-represented in the sample.

For example, suppose you were selecting students from a school with the aim of determining the mean height of all the students in the school. You would not choose a sample consisting entirely of first-year students because you would expect that such a sample would underestimate the true mean height. When a sampling method does over-represent or under-represent a feature of the population it is said to be **biased**. Any good selection method should try to reduce the chance of bias as far as possible.

The most common approach to the task of avoiding bias is to select a **random sample**.

> A random sample of size n is a sample chosen in such a way
> that each possible group of size n which could be taken
> from the population has the same chance of being picked.

4.2 Choosing a random sample

In order to select a random sample you need a list of all the members of the population. This list is called a **sampling frame**.

Suppose, for example, that the population consists of 450 different objects and that you want to take a random sample of size 20. You would assign each member of the population a unique identification number between 1 and 450 inclusive. You would then write each of the numbers between 1 and 450 on separate tickets and put them into a large hat (or other container). To obtain a random sample of size 20 you draw out 20 tickets from the hat, one after another, just as you would in a lottery competition. You must ensure that the tickets are thoroughly shuffled between each selection and you must select the tickets without replacement.

The method described above is rather time-consuming if the size of the population is large and so it is more common to adopt alternative methods which use random number tables or the random number facility on a calculator or computer.

The random number facility on a calculator provides a decimal number between 0 and 1 given to a fixed number of decimal places. The number of decimal places given varies depending on the make of calculator. Some makes give the decimal with 10 digits after the decimal point. To get your calculator to produce this random decimal you will probably need to look for the command [rand] or [rnd] on your calculator. Each time you give the calculator this command it will produce a 10 digit decimal number between 0.00000 00000 and 0.99999 99999 inclusive. A different 10 digit decimal appears each time you use [rand]. Strictly speaking, the decimals produced in this way are not random because the calculator uses a system of rules to produce each decimal and if you knew the rules you could predict each decimal number. This means that the digits produced are not really independent of one another. For a truly random sequence of digits, the process which produces the digits should ensure that each digit in the sequence is equally likely to be any of the digits 0, 1, 2, ... , 9, independently of any of the other digits. Nevertheless the sequences produced in this way do have some of the properties of random digits and they are usually considered to be acceptable as an alternative to a truly random sequence of digits.

In this book, random numbers will be written in blocks of five digits.

A **random number table** is a list of integers 0, 1, 2, ... , 9 which is produced so that each digit is equally likely to appear in any position in the list. This list of digits has been taken from a set of random number tables.

24359	74025	90831	88610
14668	78292	51470	17505
40580	96418	73381	23112

Although the tables have been arranged in blocks of five digits, you can regard the digits as forming a single long list. The procedure for selecting a random sample is given below. The population size is 450 and each member of the population has a unique identification number between 1 and 450 inclusive.

- Choose some starting point in the tables at random (by using a dice or by some similar method). The digit in bold font shows the randomly chosen starting point.

24359	74025	90831	88610
14668	78292	5**1**470	17505
40580	96418	73381	23112

- The population size is a three-digit number so read off the first three digits starting with the bold-faced digit, obtaining the number 147. This means that the first object in your sample is the object whose identification number is 147.
- Since the sample is to be of size 20, you need to choose 19 more objects from the population. Take the next block of three digits, 017, and select the object numbered 17. The third block is 505; this does not correspond to the identification number of any member of the population, so ignore it and move on to the next block, 405. Continue in this way, selecting blocks of three digits until you have a sample of 20 objects. If a three-digit block occurs more than once, ignore any later occurrence and move to the next block.

You may think that it is wasteful to discard all the blocks of digits which give numbers greater than 450. You can avoid this wastage by allocating a second block of three digits to each member of the population. Then 900 of the 1000 possible three-digit combinations would correspond to objects in your population. You cannot allocate the remaining 100 blocks to members of the population because that would give some members of the population a greater chance of being chosen than others. For a random sample every member of the population must have an equal chance of being selected.

There are many ways in which you can allocate two blocks to each member of the population. Ideally you want a way which makes it easy to identify which member of the population has been selected. Table 4.1 shows one possible way.

Number identifying population member	Random digit blocks
1	001 002
2	003 004
⋮	⋮
450	899 900

Table 4.1. Allocating random numbers to a population.

It is easy to identify which member of the population has been selected using the system in Table 4.1. For an even-numbered block you halve its value to find the appropriate member of the population and for an odd-numbered block you add one and then halve the number. For example, block 420 corresponds to object number $\frac{1}{2} \times 420 = 210$ and block 555 corresponds to object number $\frac{1}{2} \times (555 + 1) = 278$.

If the population size is greater than 1000 but less than 10 000 you can adapt the sampling method described by taking blocks containing four random digits rather than three.

It is easy to adjust the method if you are using the random number facility on a calculator. Suppose, for example, you obtain the random decimal 0.33936 62525. You can ignore the decimal point and treat the 10 digits as 10 random digits 33936 62525. After that you can use the method which was given for use with random number tables.

In some textbooks you may see an alternative method suggested for using the random number facility on a calculator. The instructions below summarise this method for a population of size N.

- Select a random decimal, r, on your calculator.
- Multiply r by N.
- Take the whole number part of this answer and add 1.

This process is sometimes expressed by giving the formula

$$\text{identification number of the selected member of the population} = \text{int}(N \times r) + 1,$$

where $\text{int } x$ is a function whose output is the greatest integer less than or equal to x.

So, for example, with $N = 450$ and $r = 0.04399\,19875$,

$$\text{int}(N \times r) + 1 = \text{int}(450 \times 0.04399\,19875) + 1$$
$$= \text{int}(19.79\ldots) + 1 = 19 + 1 = 20.$$

This method clearly does give every member of the population a chance of being selected but is each member equally likely to be chosen? Strictly, the answer to this question is 'No', for the following reason.

There are 10^{10} different random decimals with 10 digits and by the method above each one generates an integer between 1 and 450 inclusive. However, 450 does not divide exactly into 10^{10}, so some of the values between 1 and 450 must be more likely to occur than others. In practice the differences between how often each digit occurs are small and this method is often used as a quick practical method of generating a random sample.

You should realise that even when you have taken a random sample you are not guaranteed that your sample will be representative. For example, upon selecting a random sample of 10 students from a school with the aim of estimating their mean height you might find that every member of the sample *is* a first-year student. It is not very likely but it is certainly possible. The method of random sampling does not guarantee that the chosen sample is representative. What is guaranteed is that the *method of selection* is free from bias.

You should also realise that it is not always possible to select a random sample. You will recall that you need a sampling frame before you can select a random sample from a population. Sometimes this is not possible. For example, suppose that your population is all men who are colour blind. No list is available for this population so a random sample cannot be taken. Other, non-random methods have been devised to take sensible samples in such cases.

Example 4.2.1
An insurance company receives a large number of claims for storm damage. Following a spell of particularly stormy weather 42 claims are received on a single day. Sufficient staff are available to investigate only 6 of these. The claims are numbered 01 to 42 and several suggestions are made as to how the sample of 6 should be selected. Comment on each of the following methods, including an explanation of whether it would yield a random sample or not. In each case 6 claims are required.

(a) Choose the six largest claims.

(b) Select two-digit random numbers, ignoring 00 and any number greater than 42. When six distinct numbers have been obtained choose the corresponding claims.

(c) Select two-digit random numbers. Divide each one by 42, take the remainder, add 1 and choose the corresponding claims. (For example, if 44 is selected then claim number 03 would be chosen.)

(d) As part (c), but when selecting the original random numbers ignore 84 and over.

(e) Select a single digit at random, ignoring 0, 8 and 9. Choose the claim corresponding to this number and every seventh claim thereafter. (For example, if 3 is selected choose claims numbered 03, 10, 17, 24, 31, 38.)   (OCR, adapted)

(a) Choosing the six largest claims would be a good idea if the insurance company has any doubts about the claims, as the company would not want to pay out large amounts if the claims were not valid. However, this method does not give every one of the 42 claims an equal chance of being chosen so it is not a random sampling method.

(b) This method will provide a random sample. However, many of the random digit pairs will be ignored since it is quite likely that some of the pairs will be greater than 42, so the method is inefficient and wasteful.

(c) This method will give results between 1 and 42 inclusive but not all of the numbers between 1 and 42 have an equal chance of being chosen. For example, claim 1 has three possible random digit pairs which correspond to it. They are 00, 42 and 84. Claim 30 only has two random digit pairs with which it corresponds, 29 and 71. Therefore the suggested method will not provide a random sample.

(d) The objection that some of the pairs will occur more often than others is no longer valid. The method, as described, will provide a random sample.

(e) This method is usually known as a systematic sample. Each individual claim has an equal chance of being chosen but the method nevertheless does not provide a random sample. For a random sample every possible group of 6 claims needs to have an equal chance of being chosen. Using this method, however, it would be impossible, for example, to select the first 6 claims. Providing the list of claims is not arranged in any particular order this may well be a very reasonable method of selecting a sample, but it does not provide a truly random sample according to the strict definition of a random sample.

Exercise 4A

1   The makers of a lime drink wish to find out what people think of it and decide to interview a sample of shoppers. Comment briefly on the suitability of each of the following samples.

   (a) A sample of shoppers who have just bought the drink.

   (b) A sample of shoppers consisting of one person aged 20, one person aged 21, one person aged 22, and so on, up to one person aged 80.

   (c) A sample consisting of every 50th shopper.   (OCR)

2   Information about a population can be obtained from a random sample. Explain what you understand by the term random sample.

   Comment briefly on the following methods of obtaining a 'random sample' of people from a large town.

   (a) Choose random names from the town's telephone directory.

   (b) Visit every 10th house in a certain area of the town on a Wednesday morning.

   (c) Choose at random people from each postal district in the town in proportion to the population of each district.   (OCR, adapted)

3 The editor of a local newspaper wants to investigate the age distribution of the people who read the paper, and to obtain this information a random sample of the paper's readership is required. For each of the following sampling methods, give one reason why the method may be unsatisfactory.

   (a) Reporters from the paper visit various local newsagents and interview a selection of customers in the shops who buy the paper.

   (b) A form is printed in one issue of the newspaper, and readers are invited to fill in their details, cut out the form and send it (post free) to the newspaper's office.

(OCR, adapted)

4 Give a reason why the following procedure will not give a random sample of the letters of the alphabet.

   Repeatedly choose a page at random from a dictionary and take the initial letter of the first word defined on the page.

   Describe a suitable way of obtaining such a sample.   (OCR, adapted)

4.3 The sampling distribution of the mean

In order to investigate the relation between a population and samples taken from it, it is helpful to start with a simple practical situation.

Suppose that you are sampling throws of a fair cubical dice. The best way to do this randomly is to roll dice. If you want a sample of size one you would roll a single dice; if you want a sample of size two you would roll two dice, which is the same as rolling a single dice two times in sequence. For a sample of size n, you would roll n dice, or a single dice n times in sequence. To make things clear, imagine the following situation, starting with the simplest case.

Samples of size one

Suppose that you are playing a game in which you receive a prize, in $, equal to the score obtained on one dice throw. Before you start to play the game you do not know exactly what the value of your prize will be, so your prize is a random variable.

For a single throw of the dice the score, X, has the probability distribution given in Table 4.2. This is illustrated in Fig. 4.3.

Value, x	1	2	3	4	5	6
$P(X = x)$	$\frac{1}{6}$	$\frac{1}{6}$	$\frac{1}{6}$	$\frac{1}{6}$	$\frac{1}{6}$	$\frac{1}{6}$

Table 4.2. Probability distribution for X, the score when a single fair dice is thrown.

The mean, μ, of the distribution of X, is equal to $3\frac{1}{2}$. You can see that this is true by considering the symmetry of the distribution.

To find the variance of X, use the formula $\sigma^2 = \text{Var}(X) = \sum x_i^2 p_i - \mu^2$ from S1 Section 8.2.

In this case

$$\sigma^2 = \text{Var}(X)$$
$$= 1^2 \times \tfrac{1}{6} + 2^2 \times \tfrac{1}{6} + 3^2 \times \tfrac{1}{6} + 4^2 \times \tfrac{1}{6}$$
$$+ 5^2 \times \tfrac{1}{6} + 6^2 \times \tfrac{1}{6} - \left(3\tfrac{1}{2}\right)^2$$
$$= \left(1^2 + 2^2 + 3^2 + 4^2 + 5^2 + 6^2\right) \times \tfrac{1}{6}$$
$$- \left(3\tfrac{1}{2}\right)^2$$
$$= \tfrac{91}{6} - \tfrac{49}{4} = \tfrac{35}{12}.$$

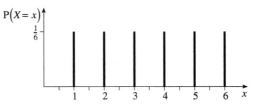

Fig. 4.3. Bar chart showing the distribution of $P(X = x)$.

Samples of size two

Suppose now that the rule of the game is that you receive a prize, in $, equal to the mean of the scores obtained on two dice throws. For example, if it turned out that you threw a 4 and then a 1 your prize would be $2.50 whereas if you scored 6 and 4 your prize would be $5. Before you start to play the game you do not know exactly what the value of your prize will be, so the value of your prize is a random variable.

If you let X_1 be the score on the first dice throw and X_2 be the score on the second dice throw, then the mean score can be expressed in terms of X_1 and X_2 as $\tfrac{1}{2}(X_1 + X_2)$. Notice that each of X_1 and X_2 has a distribution identical to the distribution of X above.

So in this example $\tfrac{1}{2}(X_1 + X_2)$ is a mean but it is also a random variable because it depends on the two separate scores, which are themselves random. The natural symbol to use for this random variable is $\overline{X}$. (Notice that you saw the symbol $\bar{x}$ before, in S1 Section 2.5, but here the symbol is given in upper case (capitals) because it is a random variable.) In this chapter, where there will be discussion about the size of the sample, it will be called $\overline{X}(2)$ as it represents the mean of two dice scores, or equivalently the mean value of a sample of size two drawn from the distribution of X.

Since $\overline{X}(2)$ is a random variable, it has a probability distribution associated with it. Table 4.4 shows all 36 equally likely possible outcomes of two fair dice throws. The entries represent the mean scores of the 36 possible pairs.

		x_2, Value of X_2					
		1	2	3	4	5	6
	1	1	$1\tfrac{1}{2}$	2	$2\tfrac{1}{2}$	3	$3\tfrac{1}{2}$
	2	$1\tfrac{1}{2}$	2	$2\tfrac{1}{2}$	3	$3\tfrac{1}{2}$	4
x_1, Value of X_1	3	2	$2\tfrac{1}{2}$	3	$3\tfrac{1}{2}$	4	$4\tfrac{1}{2}$
	4	$2\tfrac{1}{2}$	3	$3\tfrac{1}{2}$	4	$4\tfrac{1}{2}$	5
	5	3	$3\tfrac{1}{2}$	4	$4\tfrac{1}{2}$	5	$5\tfrac{1}{2}$
	6	$3\tfrac{1}{2}$	4	$4\tfrac{1}{2}$	5	$5\tfrac{1}{2}$	6

Table 4.4. Possible mean scores when two fair dice are thrown.

Of the 36 possible pairs, (x_1, x_2), there are four cases for which the mean score is $2\frac{1}{2}$. The four cases are $(1,4)$, $(2,3)$, $(3,2)$ and $(4,1)$. So the probability $P\left(\overline{X}(2) = 2\frac{1}{2}\right) = \frac{4}{36} = \frac{1}{9}$. By counting all (x_1, x_2) pairs that give the 11 possible mean scores you should be able to verify that Table 4.5 gives the probability distribution for $\overline{X}(2)$. This distribution is called the **sampling distribution of the mean**, in this case for the mean of two throws of a fair dice. The distribution is illustrated in Fig. 4.6.

Value, x	1	$1\frac{1}{2}$	2	$2\frac{1}{2}$	3	$3\frac{1}{2}$	4	$4\frac{1}{2}$	5	$5\frac{1}{2}$	6
$P\left(\overline{X}(2) = x\right)$	$\frac{1}{36}$	$\frac{2}{36}$	$\frac{3}{36}$	$\frac{4}{36}$	$\frac{5}{36}$	$\frac{6}{36}$	$\frac{5}{36}$	$\frac{4}{36}$	$\frac{3}{36}$	$\frac{2}{36}$	$\frac{1}{36}$

Table 4.5. Probability distribution of the mean score when two fair dice are thrown.

There is a connection between the mean μ, of the distribution of X and the mean of the distribution of $\overline{X}(2)$. From the symmetry of the distribution, $E\left(\overline{X}(2)\right) = 3\frac{1}{2}$, so $E\left(\overline{X}(2)\right) = \mu$.

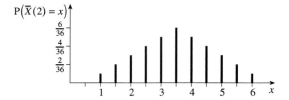

The variances of X and $\overline{X}(2)$ are also connected. The variance of $\overline{X}(2)$ is given by

Fig. 4.6. Bar chart showing the distribution of $P\left(\overline{X}(2) = x\right)$.

$$\text{Var}\left(\overline{X}(2)\right) = 1^2 \times \tfrac{1}{36} + \left(1\tfrac{1}{2}\right)^2 \times \tfrac{2}{36} + 2^2 \times \tfrac{3}{36} + \ldots + 6^2 \times \tfrac{1}{36} - \left(3\tfrac{1}{2}\right)^2$$

$$= \tfrac{35}{24} \qquad \text{(after some calculation)}.$$

You can see from this result that, as $\sigma^2 = \frac{35}{12}$, $\text{Var}\left(\overline{X}(2)\right) = \frac{1}{2}\sigma^2$.

Summarising, and returning to the language of samples, for samples of size two $E\left(\overline{X}(2)\right) = \mu$ and $\text{Var}\left(\overline{X}(2)\right) = \frac{1}{2}\sigma^2$, where $E(X) = \mu$ and $\text{Var}(X) = \sigma^2$.

These results are true for *all* distributions, not just for this particular one. The results follow from the rules established in Chapter 2 for linear combinations of random variables.

$$E\left(\overline{X}(2)\right) = E\left(\tfrac{1}{2}(X_1 + X_2)\right) = E\left(\tfrac{1}{2}X_1 + \tfrac{1}{2}X_2\right)$$

$$= \tfrac{1}{2}E(X_1) + \tfrac{1}{2}E(X_2) \qquad \text{(from Equation 2.7)}$$

$$= \tfrac{1}{2}\mu + \tfrac{1}{2}\mu$$

$$= \mu.$$

and $\quad \text{Var}\left(\overline{X}(2)\right) = \text{Var}\left(\tfrac{1}{2}(X_1 + X_2)\right) = \text{Var}\left(\tfrac{1}{2}X_1 + \tfrac{1}{2}X_2\right)$

$$= \left(\tfrac{1}{2}\right)^2 \text{Var}(X_1) + \left(\tfrac{1}{2}\right)^2 \text{Var}(X_2) \qquad \text{(from Equation 2.8)}$$

$$= \tfrac{1}{4}\sigma^2 + \tfrac{1}{4}\sigma^2$$

$$= \tfrac{1}{2}\sigma^2.$$

Samples of size three

Perhaps you can now predict the values of $E(\overline{X}(3))$ and $Var(\overline{X}(3))$ in terms of μ and σ^2 for samples of size three, or, equivalently, three dice throws. Let $\overline{X}(3)$ denote the mean score on three dice throws.

Let X_1, X_2 and X_3 be the scores on the first, second and third throws respectively. Then the mean $\overline{X}(3) = \frac{1}{3}(X_1 + X_2 + X_3)$.

Again $\overline{X}(3)$ is a random variable because X_1, X_2 and X_3 are random variables.

One possible value for $\overline{X}(3)$ is $2\frac{1}{3}$.

There are four different combinations of scores which give a mean score of $2\frac{1}{3}$: they are $(1,1,5)$, $(1,2,4)$, $(1,3,3)$ and $(2,2,3)$. The combination $(1,1,5)$ can occur in any of 3 different orders: $(1,1,5)$, $(1,5,1)$, $(5,1,1)$; similarly $(1,2,4)$ can occur in 6 orders: $(1,2,4)$, $(1,4,2)$, $(2,1,4)$, $(2,4,1)$, $(4,1,2)$, $(4,2,1)$; $(1,3,3)$ can occur in 3 orders: $(1,3,3)$, $(3,1,3)$, $(3,3,1)$; and $(2,2,3)$ can occur in 3 orders: $(2,2,3)$, $(2,3,2)$, $(3,2,2)$.

From these results, you can see that

$$P(\overline{X}(3) = 2\frac{1}{3}) = \frac{3}{216} + \frac{6}{216} + \frac{3}{216} + \frac{3}{216} = \frac{15}{216}.$$

By considering all (x_1, x_2, x_3) triples that give each of the 16 possible mean scores, you can derive the probability distribution of $\overline{X}(3)$. This is given in Table 4.7. This is the sampling distribution of the mean for samples of 3 throws of a fair dice.

Value, x	$P(\overline{X}(3) = x)$	Value, x	$P(\overline{X}(3) = x)$
1	$\frac{1}{216}$	$3\frac{2}{3}$	$\frac{27}{216}$
$1\frac{1}{3}$	$\frac{3}{216}$	4	$\frac{25}{216}$
$1\frac{2}{3}$	$\frac{6}{216}$	$4\frac{1}{3}$	$\frac{21}{216}$
2	$\frac{10}{216}$	$4\frac{2}{3}$	$\frac{15}{216}$
$2\frac{1}{3}$	$\frac{15}{216}$	5	$\frac{10}{216}$
$2\frac{2}{3}$	$\frac{21}{216}$	$5\frac{1}{3}$	$\frac{6}{216}$
3	$\frac{25}{216}$	$5\frac{2}{3}$	$\frac{3}{216}$
$3\frac{1}{3}$	$\frac{27}{216}$	6	$\frac{1}{216}$

Table 4.7. Probability distribution of the mean score when three fair dice are thrown.

This distribution is illustrated in Fig. 4.8.

From the symmetry of the distribution, the expected value, $E(\overline{X}(3))$, is again $3\frac{1}{2}$.

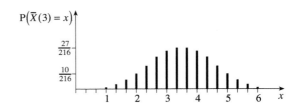

Fig. 4.8. Bar chart showing the distribution of $P(\overline{X}(3) = x)$.

The variance is given by

$$\text{Var}(\overline{X}(3)) = 1^2 \times \tfrac{1}{216} + \left(1\tfrac{1}{3}\right)^2 \times \tfrac{3}{216} + \left(1\tfrac{2}{3}\right)^2 \times \tfrac{6}{216} + \ldots + 6^2 \times \tfrac{1}{216} - \left(3\tfrac{1}{2}\right)^2$$

$$= \left(1 + \tfrac{16}{9} \times 3 + \tfrac{25}{9} \times 6 + \ldots + 36\right) \times \tfrac{1}{216} - \left(3\tfrac{1}{2}\right)^2$$

$$= \tfrac{35}{36} \quad \text{(after some calculation).}$$

So this time $\text{Var}(\overline{X}(3)) = \tfrac{1}{3}\sigma^2$.

Try to prove this for all distributions by using Equations 2.7 and 2.8.

The results for samples of size 2 and size 3 can be generalised for a sample of size n, to give $\text{E}(\overline{X}(n)) = \mu$ and $\text{Var}(\overline{X}(n)) = \dfrac{\sigma^2}{n}$. These result can be proved as follows.

$$\text{E}(\overline{X}(n)) = \text{E}\left(\frac{1}{n}(X_1 + X_2 + \ldots + X_n)\right)$$

$$= \frac{1}{n}\text{E}(X_1) + \frac{1}{n}\text{E}(X_2) + \ldots + \frac{1}{n}\text{E}(X_n) \quad \text{(from Equation 2.7)}$$

$$= \frac{1}{n}\mu + \frac{1}{n}\mu + \ldots + \frac{1}{n}\mu$$

$$= n \times \frac{1}{n}\mu = \mu.$$

$$\text{Var}(\overline{X}(n)) = \text{Var}\left(\frac{1}{n}(X_1 + X_2 + \ldots + X_n)\right)$$

$$= \frac{1}{n^2}\text{Var}(X_1) + \frac{1}{n^2}\text{Var}(X_2) + \ldots + \frac{1}{n^2}\text{Var}(X_n) \quad \text{(from Equation 2.8)}$$

$$= \frac{1}{n^2}\sigma^2 + \frac{1}{n^2}\sigma^2 + \ldots + \frac{1}{n^2}\sigma^2$$

$$= n \times \frac{1}{n^2}\sigma^2 = \frac{\sigma^2}{n}.$$

If a random sample consists of n observations of a random variable X, and the mean $\overline{X}$ is found, then $\text{E}(\overline{X}(n)) = \mu$ and $\text{Var}(\overline{X}(n)) = \dfrac{\sigma^2}{n}$, where $\mu = \text{E}(X)$ and $\sigma^2 = \text{Var}(X)$.

Example 4.3.1

A biased coin for which the probability of turning up heads is $\tfrac{2}{3}$ is spun 20 times. Let $\overline{X}$ denote the mean number of heads per spin. Calculate $\text{E}(\overline{X})$ and $\text{Var}(\overline{X})$.

The random variable X in this case represents the number of heads in one spin of the coin. The distribution of X is given in the table.

Value	0	1
$P(X = x)$	$\frac{1}{3}$	$\frac{2}{3}$

The mean is $\mu = E(X) = 0 \times \frac{1}{3} + 1 \times \frac{2}{3} = \frac{2}{3}$ and the variance is given by

$$\sigma^2 = \text{Var}(X) = 0^2 \times \frac{1}{3} + 1^2 \times \frac{2}{3} - \left(\frac{2}{3}\right)^2 = \frac{2}{9}.$$

Using the result in the box,

$$E\left(\overline{X}(20)\right) = \mu = \frac{2}{3} \quad \text{and} \quad \text{Var}\left(\overline{X}(20)\right) = \frac{\sigma^2}{20} = \frac{1}{20} \times \frac{2}{9} = \frac{1}{90}.$$

4.4 The central limit theorem

Fig. 4.9 compares the bar charts which were obtained in the previous section for the distribution of the throw on a dice and the mean scores when 2 and 3 dice are thrown.

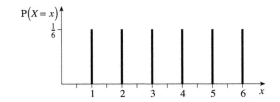

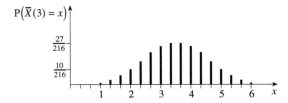

Fig. 4.9. Figs. 4.3, 4.6 and 4.8 redrawn to compare them.

The distribution for a single throw is uniform whereas those for $\overline{X}(2)$ and for $\overline{X}(3)$ are not. The distribution for $\overline{X}(2)$ has a 'peak' in the middle and the values in the centre of the interval, 3, $3\frac{1}{2}$ and 4, are more likely to occur than the other values. When three dice are thrown the central values occur even more often. In fact, the bar chart for $\overline{X}(3)$ looks very similar to the graph of the normal distribution.

As n gets larger the distribution of $\overline{X}(n)$ gets more and more like a normal distribution. Fig. 4.10 shows the result of a simulation using a computer spreadsheet for a sample of size 50.

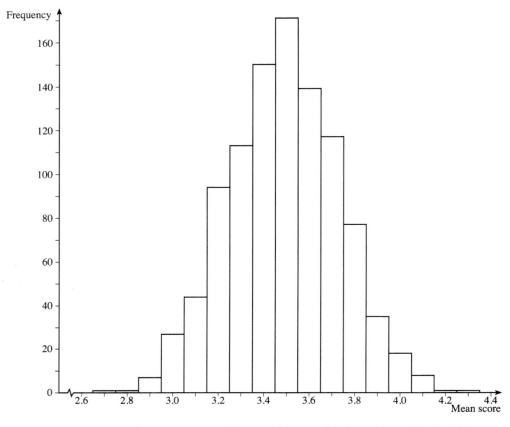

Fig. 4.10. Diagram showing the frequency distribution of the mean of 50 throws of a fair dice for 1000 simulations on a spreadsheet.

If you have access to a spreadsheet, you can try such a simulation for yourself. The shape of the diagram becomes more clearly normal as the number of simulations is increased.

So, from the examples you have considered, it would seem reasonable to conclude that the distribution of $\overline{X}(n)$ is approximately normal for large values of n. From the previous section, $\mathrm{E}\big(\overline{X}(n)\big) = \mu$ and $\mathrm{Var}\big(\overline{X}(n)\big) = \dfrac{\sigma^2}{n}$.

These results are special cases of a more general theorem called the **central limit theorem.**

The central limit theorem   For any sequence of independent identically distributed random variables X_1, X_2, ... , X_n with finite mean μ and non-zero variance σ^2, then, provided n is sufficiently large, $\overline{X}$ has approximately a normal distribution with mean μ and variance $\dfrac{\sigma^2}{n}$, where $\overline{X} = \dfrac{X_1 + X_2 + ... + X_n}{n}$.

In symbols, $\overline{X} \sim \mathrm{N}\!\left(\mu, \dfrac{\sigma^2}{n}\right)$.

Notice that the sample size n has been omitted from $\overline{X}$ because it is usually clear from the context how large the sample is.

You may have been wondering just how large n has to be for the normal distribution to be reasonably close to the distribution of $\overline{X}$. Unfortunately there is no simple answer. It depends on the distribution of X itself. If the distribution of X has some of the features of a normal distribution then n will probably not need to be very large at all. On the other hand, if the distribution of X is very skewed then n might need to be quite large. In solving problems you can usually judge whether it is appropriate to invoke the central limit theorem. A sensible rule of thumb is to apply the theorem when n is greater than about 30.

This theorem is a fundamental result in the theory of statistics and it explains why the normal distribution is so widely studied. The essential point is that it does not matter what distribution X_1, X_2, ... , X_n have individually: as long as they all have the same distribution and are independent of one another, the distribution of the mean $\overline{X}$ will be approximately normal as long as n is sufficiently large. The central limit theorem is too complicated to prove rigorously here; however, the previous discussion of the mean scores on one, two and three dice shows that the theorem is reasonable. The following examples show how the theorem applies to other random variables.

Example 4.4.1
A continuous random variable, X, has a probability density function, $f(x)$, given by

$$f(x) = \begin{cases} 2x & \text{for } 0 \leqslant x \leqslant 1, \\ 0 & \text{otherwise.} \end{cases}$$

Find   (a) the mean, μ,   (b) the variance, σ^2, of this distribution.

A random sample of 100 observations is taken from this distribution, and the mean $\overline{X}$ is found. Calculate the probability $P(\overline{X} < 0.68)$.

Using the definitions of the mean and variance,

(a) $\mu = \int_0^1 x \times 2x \, dx = \int_0^1 2x^2 \, dx = \left[\frac{2}{3}x^3\right]_0^1 = \frac{2}{3}$ (from Equation 3.1).

(b) $\sigma^2 = \int_0^1 x^2 \times 2x \, dx - \left(\frac{2}{3}\right)^2 = \int_0^1 2x^3 \, dx - \left(\frac{2}{3}\right)^2$ (from Equation 3.2)

$= \left[\frac{2}{4}x^4\right]_0^1 - \left(\frac{2}{3}\right)^2 = \frac{1}{2} - \frac{4}{9} = \frac{1}{18}$.

By the central limit theorem, the distribution of $\overline{X}$ is approximately $N\left(\frac{2}{3}, \frac{1}{1800}\right)$.

Standardising, using $Z = \dfrac{\overline{X} - \frac{2}{3}}{\sqrt{\frac{1}{1800}}}$, it follows that

$$P(\overline{X} < 0.68) \approx P\left(Z < \frac{0.68 - \frac{2}{3}}{\sqrt{\frac{1}{1800}}}\right) = P(Z < 0.565\ 68...) = 0.7143$$

$$= 0.714, \text{ correct to 3 significant figures.}$$

Example 4.4.2

Forty students each threw a fair cubical dice 12 times. Each student then recorded the number of times that a six occurred in their own 12 throws. The students' lecturer then calculated the mean number of sixes obtained per student. Find the probability that this mean was over 2.2.

Let X_i be the number of sixes obtained by student i for $i = 1, 2, \ldots, 40$.

Each X_i satisfies the conditions for a binomial distribution to apply. The parameters of the binomial distribution in this case are $n = 12$ and $p = \frac{1}{6}$.

Recall from S1 Section 8.3 that for a random variable X which has a binomial distribution, $E(X) = np$ and $Var(X) = npq$. So, in this case, $E(X) = 12 \times \frac{1}{6} = 2$ and $Var(X) = 12 \times \frac{1}{6} \times \frac{5}{6} = \frac{5}{3}$.

Using the central limit theorem, $\overline{X} \sim N\left(2, \dfrac{\frac{5}{3}}{40}\right) = N\left(2, \frac{1}{24}\right)$, approximately.

$\overline{X}$ represents the mean of 40 binomial variables, so it can be written in terms of $X_1, X_2, \ldots, X_{40}$ as $\overline{X} = \frac{1}{40}(X_1 + X_2 + \ldots + X_{40})$.

You want to find $P(\overline{X} > 2.2)$. This can be written in terms of T, the total of the 40 variables, where $T = X_1 + X_2 + \ldots + X_{40}$.

$P(\overline{X} > 2.2)$ is equivalent to $P(T > 40 \times 2.2)$, which is $P(T > 88)$. However, T is the total number of sixes gained in 480 throws of a fair dice, so $T \sim B\left(480, \frac{1}{6}\right)$.

From S1 Section 9.7, when you use a normal distribution to approximate to a binomial distribution, you need a continuity correction, so $P(T > 88)$ is approximately equal to $P(V > 88 + 0.5)$, where V is the appropriate normal approximation.

Expressing this in terms of $\overline{X}$, you want $P\left(\overline{X} > 2.2 + \frac{1}{80}\right)$.

So here, when it is applied to the mean, $\overline{X}$, of a set of n discrete variables, the continuity correction is $\dfrac{1}{2n}$, rather than $\frac{1}{2}$.

Standardising $\overline{X}$ using the usual equation gives $Z = \dfrac{\overline{X} - 2}{\sqrt{\frac{1}{24}}}$, so $Z \sim N(0,1)$.

Then $P\left(\overline{X} > 2.2 + \frac{1}{80}\right) = P\left(Z > 0.2125\sqrt{24}\right) = P(Z > 1.041\ldots)$

$$= 1 - \Phi(1.041\ldots) = 0.1490\ldots$$

$$= 0.149, \text{ correct to 3 significant figures.}$$

4.5 Practical activities

1 Playing cards   Take a pack of playing cards and discard the picture cards, that is the jack, queen and king, of each suit. Shuffle the cards and deal out 10 of them. Calculate the mean score by adding together the 10 face values and dividing by 10. Put the 10 cards back in the pack and shuffle again. Deal out a further 10 cards and calculate the mean score again. Repeat this to give 100 observations, each of which is the mean score of 10 cards. Calculate the mean and the variance of these scores. Compare these values with the ones you would expect from the theoretical results.

2 Central limit

This activity requires access to a computer.

(a) Use a spreadsheet to produce a random number, r, between 0 and 1 in one cell.

(b) In a second cell calculate $n = \frac{1}{2}\left(1 + \sqrt[3]{2r-1}\right)$.

(c) Repeat this in a large number of pairs of cells (say 500) to get 500 observations of n.

(d) Use your spreadsheet to draw a histogram of the values of n. This should give you a U-shaped distribution.

(e) Using a new spreadsheet calculate 50 values of n using the same formula as in part (b) and find their mean. Repeat this to get at least 100 means of 50 values of n.

(f) Draw a histogram of these 100 means. You should be able to see that the distribution of the means is not U-shaped like the original distribution. The distribution of the means should more nearly resemble a normal distribution.

4.6 The distribution of $\overline{X}$ when $X_1, X_2, \ldots, X_n$ have a $N(\mu, \sigma^2)$ distribution

The central limit theorem states that the distribution of $\overline{X}$ is approximately $N\left(\mu, \dfrac{\sigma^2}{n}\right)$ when $X_1, X_2, \ldots, X_n$ are independent identically distributed random variables with finite mean μ and finite non-zero variance σ^2. It is necessary for the value of n to be quite large for this approximation to be reasonable. However, if the distributions of $X_1, X_2, \ldots, X_n$ are independent $N(\mu, \sigma^2)$ variables, then the distribution of $\overline{X}$ is **exactly** $N\left(\mu, \dfrac{\sigma^2}{n}\right)$ no matter what size n is. This result follows from the fact, stated in Chapter 2, that linear combinations of normal variables have a normal distribution. The result is used in Example 4.6.1.

Example 4.6.1

(a) The mass of a randomly chosen 15-year-old male student at a large secondary school may be modelled by a normal distribution with mean 55 kg and standard deviation 2.2 kg. Four students are chosen at random from this group. Calculate the probability that the mean mass of the four students is
(i) less than 58 kg,   (ii) between 52 kg and 57.5 kg.

(b) A second sample of size n is chosen from the 15-year-old male students. How large does n have to be for there to be at most a 2% chance that the mean mass of the sample differs from the mean mass of the population by more than 0.6 kg?

Let M_1, M_2, M_3 and M_4 be the masses of four randomly chosen students. Then

$$M_i \sim N(55, 2.2^2) \text{ for } i = 1, 2, 3, 4.$$

Therefore $\overline{M} \sim N\left(55, \dfrac{2.2^2}{4}\right) = N(55, 1.21)$.

Note that, although n is very small, as the distribution for each M_i is normal the mean of $\overline{M}$ is normal.

Standardise by letting $Z = \dfrac{\overline{M} - 55}{1.1}$, then $Z \sim N(0,1)$.

(a) (i) $P(\overline{M} < 58) = P\left(Z < \dfrac{58 - 55}{1.1}\right) = P(Z < 2.727\ldots)$

$\qquad\qquad = \Phi(2.727\ldots) = 0.9968 = 0.997$, correct to 3 significant figures.

(ii) $P(52 < \overline{M} < 57.5) = P\left(\dfrac{52 - 55}{1.1} < Z < \dfrac{57.5 - 55}{1.1}\right) = P(-2.727\ldots < Z < 2.272\ldots)$

$\qquad\qquad\qquad = \Phi(2.272\ldots) - \Phi(-2.727\ldots)$

$\qquad\qquad\qquad = 0.9884 - (1 - 0.9968)$

$\qquad\qquad\qquad = 0.985$, correct to 3 significant figures.

(b) Let M_1, M_2, $\ldots$, M_n be the masses of n randomly chosen students.

Then

$$\overline{M} \sim N\left(55, \dfrac{2.2^2}{n}\right)$$

Then the minimum value of n is required for which

$$P(54.4 \leqslant \overline{M} \leqslant 55.6) \geqslant 0.98.$$

Standardising this gives

$$P\left(\dfrac{-0.6}{2.2/\sqrt{n}} \leqslant Z \leqslant \dfrac{0.6}{2.2/\sqrt{n}}\right) \geqslant 0.98.$$

Expressing this in terms of the cumulative distribution function gives

$$\Phi\left(\dfrac{0.6}{2.2/\sqrt{n}}\right) - \Phi\left(\dfrac{-0.6}{2.2/\sqrt{n}}\right) \geqslant 0.98.$$

Therefore $\Phi\left(\dfrac{0.6}{2.2/\sqrt{n}}\right) - \left(1 - \Phi\left(\dfrac{0.6}{2.2/\sqrt{n}}\right)\right) \geqslant 0.98$.

Solving this gives $\Phi\left(\dfrac{0.6}{2.2/\sqrt{n}}\right) \geqslant 0.99$.

Taking the inverse function $\dfrac{0.6}{2.2/\sqrt{n}} \geqslant \Phi^{-1}(0.99)$.

From the tables, this gives $\dfrac{0.6}{2.2/\sqrt{n}} \geqslant 2.326$.

Rearranging this equation gives $\sqrt{n} \geqslant 8.528\ldots$, so $n \geqslant 72.7\ldots$.

Since n must be an integer, it has to be greater than or equal to 73. So the sample needs to contain at least 73 students for its mean mass to approximate the population's mean mass with the accuracy and certainty desired.

Exercise 4B

1 The random variable X has mean μ and variance σ^2. State whether or not each of the following statements relating to the distribution of the mean $\overline{X}$ of a random sample of n observations of X is true. Correct any false statement.

   (a) The central limit theorem states that $\overline{X}$ has a normal distribution for any distribution of X.

   (b) $\mathrm{E}(\overline{X}) = \mu$ and $\mathrm{Var}(\overline{X}) = \dfrac{\sigma^2}{n}$ are true for any distribution of X and any value of n.

   (c) The central limit theorem states that the sample is normally distributed for large values of n.

   (d) If $X \sim \mathrm{N}(\mu, \sigma^2)$ then $\overline{X} \sim \mathrm{N}\left(\mu, \dfrac{\sigma^2}{n}\right)$ only for large values of n.

2 Random samples of three are drawn from a population of beetles whose lengths have a normal distribution with mean 2.4 cm and standard deviation 0.36 cm. The mean length $\overline{X}$ is calculated for each sample.

   (a) State the distribution of $\overline{X}$, giving the values of its parameters.

   (b) Find $\mathrm{P}(\overline{X} > 2.5)$.

   State which of the numerical values above, if any, depend on the central limit theorem.

3 An unbiased dice is thrown once. Write down the probability distribution of the score X and show that $\mathrm{Var}(X) = \frac{35}{12}$.

   The same dice is thrown 70 times.

   (a) Find the probability that the mean score is less than 3.3.

   (b) Find the probability that the total score exceeds 260.

4 The masses of kilogram bags of flour produced in a factory have a normal distribution with mean 1.005 kg and standard deviation 0.0082 kg. A shelf in a store is loaded with 22 of these bags, assumed to be a random sample.

(a) Find the probability that a randomly chosen bag has mass less than 1 kg.

(b) Find the probability that the mean mass of the 22 bags is less than 1 kg.

State, giving a reason, which of the above answers would be little changed if the distribution of masses were not normal.

5 A rectangular field is gridded into squares of side 1 m. At one time of the year the number of snails in the field can be modelled by a Poisson distribution with mean 2.25 per m^2.

(a) A random sample of 120 squares is observed and the number of snails in each square counted. Find the probability that the sample mean number of snails is at most 2.5.

(b) Show that the probability of observing at least 200 snails in a random sample of 100 grid squares is approximately 95%.

6 The random variable X has a B(40,0.3) distribution. The mean of a random sample of n observations of X is denoted by $\overline{X}$. Find

(a) $P(\overline{X} \geqslant 13)$ when $n = 49$,

(b) the smallest value of n for which $P(\overline{X} \geqslant 13) < 0.001$.

7 The proportion of faulty plastic cups made by a factory machine is 0.08. The cups, including faulty ones, are packed in boxes of 100. About 4000 cups are required for an outdoor concert and the manager orders 44 boxes. Find the probability that these boxes will provide more than 4000 perfect cups.

8 A liquid drug is marketed in phials containing a nominal 1.5 ml but the amounts can vary slightly. The volume in each phial may be modelled by a normal distribution with mean 1.55 ml and standard deviation σ ml. The phials are sold in packs of 5 randomly chosen phials. It is required that in less than $\frac{1}{2}$% of packs will the total volume of the drug be less than 7.5 ml. Find the greatest possible value of σ.

9 A goods lift can carry up to 5000 kg and is to be loaded with crates whose masses are normally distributed with mean 79.2 kg and standard deviation 5.5 kg. Show that

(a) it is highly likely that the lift can take 62 randomly selected crates without being overloaded,

(b) 65 randomly selected crates would almost certainly overload the lift.

10 A firm of caterers wishes to buy fruit juice for a wedding reception of 200 guests. They estimate that, on average, each guest will drink 45 cl of juice. The volume of juice in the bottles they buy may be assumed to have a distribution with mean 70.5 cl and standard deviation 1.2 cl. Show that if they buy 128 bottles then the caterers can be more than 95% certain that their requirements will be met.

11 The mean of a random sample of 500 observations of the random variable X, where $X \sim N(25,18)$ is denoted by $\overline{X}$. Find the value of a for which $P(\overline{X} < a) = 0.25$.

Does your answer depend on the central limit theorem?

Miscellaneous exercise 4

1 Eggs sold in a supermarket are packed in boxes of 12. For each egg, the probability that it is cracked is 0.05, independently of all other eggs. A random sample of n boxes is selected and the variance of the sample mean number of cracked eggs in a box is 0.019. Find the value of n. (OCR, adapted)

2 The mean of a random sample of n observations drawn from a $N(\mu, \sigma^2)$ distribution is denoted by $\overline{X}$. Given that $P(|\overline{X} - \mu| > 0.5\sigma) < 0.05$

    (a) find the smallest value of n,

    (b) with this value of n, find $P(\overline{X} < \mu + 0.1\sigma)$.

3 A botanist wishes to estimate the mean μ and standard deviation σ of the depth of the soil in a large rectangular field. Comment on the following methods of obtaining the sample points.

    (a) The botanist stands at a point near the centre of the field, facing a particular direction, and throws a stone over her shoulder. The sample point is where the stone lands. This is repeated, changing the direction she faces.

    (b) The botanist grids the field into metre squares and uses a table of random numbers to define a sample point at the centre of the square.

Assume that the botanist uses a suitable sampling procedure. She requires the sample mean depth to differ from μ by less than 10% of σ with probability at least $97\frac{1}{2}\%$. Find the smallest sample size that she will have to obtain.

4 The time T hours taken to repair a piece of equipment has a probability density function which may be modelled by

$$f(t) = \begin{cases} \dfrac{24}{7t^4} & 1 \leqslant t \leqslant 2, \\ 0 & \text{otherwise.} \end{cases}$$

    (a) Find $E(T)$ and $Var(T)$.

    (b) $\overline{T}$ denotes the mean of 30 randomly chosen repairs. Assuming that the central limit theorem holds, estimate $P(\overline{T} < 1.2)$.

State, giving a reason, whether your answer has little error or considerable error.

5 A machine is set to produce ball-bearings with mean diameter 1.2 cm. Each day a random sample of 50 ball-bearings is selected and the diameters accurately measured. If the sample mean diameter lies outside the range 1.18 cm to 1.22 cm then it will be taken as evidence that the mean diameter of the ball-bearings produced is not 1.2 cm. The machine will then be stopped and adjustments made to it. Assuming that the diameters have standard deviation 0.075 cm, find the probability that

    (a) the machine is stopped unnecessarily,

    (b) the machine is not stopped when the mean diameter of the ball-bearings is 1.15 cm.

6 The number of night calls to a fire station serving a small town can be modelled by a Poisson distribution with mean 2.7 calls per night.

(a) State the expectation and variance of the mean number of night calls over a period of n nights.

(b) Estimate the probability that during a given year of 365 days the total number of night calls will exceed 1050.

7 It may be assumed that the breaking strength of paving slabs in public areas is normally distributed with mean 50 units and standard deviation 8 units. Random samples of n paving slabs are taken and the mean breaking strength is denoted by $\overline{X}$.

(a) State the distribution of $\overline{X}$, giving its mean and variance.

(b) Find the probability that $\overline{X}$ exceeds 54 in the case $n = 25$.

(c) Find the smallest sample size if the probability that $\overline{X}$ exceeds 54 units is less than 0.01.

If it were not known that the distribution of breaking strengths is normal, what can be said about the form of the distribution of $\overline{X}$ in the case when n is large and also in the case when n is small?   (OCR, adapted)

8 A random sample of $2n$ observations is taken of the random variable $X \sim \mathrm{B}(n,p)$ and $p > 0.5$. The sample mean is denoted by $\overline{X}$. It is given that $\mathrm{Var}\left(\overline{X}\right) = 0.08$ and $\mathrm{E}\left(\overline{X}\right) = 64$.

(a) Find the values of p and n.

(b) Find $\mathrm{P}\left(\overline{X} > 64.5\right)$.

9 The life of Powerlong batteries, sold in packs of 6, may be assumed to have a normal distribution with mean 32 hours and standard deviation σ hours. Find the value of σ so that for one box in 100 (on average) the mean life of the batteries is less than 30 hours.

10 The number of tickets sold each day at a city railway station has mean 512 and variance 1600. For a randomly chosen period of 60 days, find the probability that the total number of tickets sold is less than 30 000.

11 The mean of 64 observations of a random variable X, where $\mathrm{E}(X) = 9$ and $\mathrm{Var}(X) = 4$, is denoted by $\overline{X}$. Find limits within which $\overline{X}$ lies with probability 0.96.

Chebychev's inequality states that, if Y has any distribution with mean μ and variance σ^2, then $\mathrm{P}(|Y - \mu| < k\sigma) \geqslant 1 - \dfrac{1}{k^2}$ for $k \geqslant 1$.

Use Chebychev's inequality to find another pair of limits for $\overline{X}$. Comment on these limits in relation to those already found.

5 Estimation

This chapter looks at what you can deduce about a population from a sample. When you have completed it you should

- understand the term 'unbiased' with reference to an estimator of the mean or variance of a sample and be able to calculate unbiased estimates of the population mean and variance from a sample
- be able to determine a confidence interval for a population mean in the case when the population is normally distributed or where a large sample is used
- be able to determine, from a large sample, an approximate confidence interval for a population proportion.

5.1 Unbiased estimates

One of the main aims in the study of statistics is to be able to estimate the value of population parameters without going to the trouble of taking a census.

Consider the example of a hospital administrator who is concerned that the hospital sometimes does not have enough beds to cater for all of its patients. She wants to estimate μ, the mean time for which a patient stays in a hospital bed during a particular week. She takes a random sample of the patients who are admitted during that week and records the length of time, in days, for which each patient in the sample stays in hospital. She must now consider how to use the data that she has collected in order to estimate μ. For example, suppose she has collected the times for 10 patients. The results might be something like

$$7, 5, 5, 9, 3, 11, 6, 4, 2, 20.$$

What calculation should she carry out on these values to estimate μ? You may think that the answer is obvious. It makes sense to estimate μ by taking the mean of the sample. So in this case the estimate of μ would be $\frac{72}{10} = 7.2$. But is this the only sensible estimate that can be made from these data? What would be wrong, for instance, with taking the first time, 7 days, as an estimate of μ? It may be that the real mean, μ, is nearer to 7 than to 7.2. It is clear that you need a strategy for deciding which method is better.

To make a decision you need to think in detail about what the administrator is trying to do. Imagine that she has not yet taken her sample of 10 times. So at present the times are random variables because it is not certain which patients she will choose. Suppose that you call these random variables $X_1, X_2, \ldots, X_{10}$. Two possible methods for estimating μ have been suggested.

Method 1   Use the mean of the 10 sample values. Call this value M_1. Then $M_1 = \frac{1}{10}(X_1 + X_2 + \ldots + X_{10})$. Recall that this may also be called $\overline{X}$.

Method 2   Use the first value selected. Call this value M_2. So $M_2 = X_1$.

The values obtained in these ways are themselves random variables and are usually called **estimators**. The value that an estimator gives for a particular sample is called an **estimate**.

It should be clear that $X_2, X_3, \ldots, X_{10}$ all have the same distribution as X_1, so each variable also has the expected value μ. You saw in Section 4.3 that if $X_1, X_2, \ldots, X_{10}$ is a set of random variables with the same mean, μ, then $E(\overline{X}) = \mu$.

Therefore, for both of the suggested methods, $E(M_1) = E(M_2) = \mu$.

What exactly does this statement tell you? It says that, if you took all possible samples of size 10 and calculated the value of M_1 for each sample, then the mean of these values would be equal to μ, the quantity that the administrator is trying to estimate. The same is true of M_2. This means that both estimators are **unbiased**.

> For an unknown population parameter α, the estimator
> M is unbiased if $E(M) = \alpha$, where the expected value
> is taken over all possible samples of a given size.

Fig. 5.1 represents a way of understanding the idea of unbiasedness. The line is the real number line and the crosses represent the values which an estimator gives for each sample of a population. If the estimator is

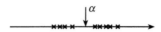

Fig. 5.1. Representation of unbiasedness.

unbiased then the mean value of all the points represented by the crosses will be the population parameter α.

Clearly you want an estimator to be unbiased because when this is the case the parameter of interest, α, is in effect the 'focus' of all the possible values that the estimator can take. Of course, this is only of theoretical importance here because the administrator has only one sample, and since both estimators are unbiased she needs some other means of deciding which estimator is the better one.

Suppose that the variance of X_1 is denoted by σ^2. Then $X_2, X_3, \ldots, X_{10}$ also have variance σ^2.

If the number of patients in the hospital during the week is a large number then you can assume that the distributions of $X_1, X_2, \ldots, X_{10}$ are independent. You saw in Section 4.3 that under these conditions $\text{Var}(\overline{X}) = \frac{1}{10}\sigma^2$.

This means that the estimates found from $M_1 = \overline{X}$ will on average be closer to μ than those found from $M_2 = X_1$. Fig. 5.2 shows a comparison between the two estimators.

Fig. 5.2. Closer estimates are found by larger samples.

You can see that the crosses are clustered more closely around μ in the first case. This confirms what you may have guessed from the start: it is better to use the estimator $M_1 = \overline{X}$ because it uses all of the information in the sample whereas using only $M_2 = X_1$ discards 90% of the information gained by taking the sample.

In general it can be shown that $\overline{X}$ is always an unbiased estimator of μ. In other words, a sensible method of estimating the population mean is to take a random sample and use its mean as the estimate.

So is this true of all population parameters? To estimate the median of a population would you take the median of a random sample, or to estimate the mode of a population would you take the mode of a random sample? The previous example would seem to suggest that this is a sensible procedure, but consider the following example.

Example 5.1.1

A jar contains a very large number of small discs, 90% of which have the number 0 written on them and the remaining 10% of which have the number 1 on them. Luc does not know what numbers are written on the discs. He takes a random sample of two discs from the jar. Luc wishes to estimate the largest number written on any disc in the jar. He decides that his method of estimating the maximum will be to take the larger of the numbers on the two discs in his sample. If both discs have the same number written on them then he takes that value as his estimate of the maximum. Decide whether the proposed method gives an unbiased estimator of the population maximum.

Let the maximum value in the sample be M.

Luc's method is to take the sample maximum, M, as an estimator of the population maximum. Table 5.3 gives all possible samples of size two and the value of M for each sample together with the probability of that sample being chosen.

Since the jar is said to contain a large number of discs it is reasonable to suppose that the chances of selecting a 0 or 1 remain constant for each selection.

Sample values	m, Value of M	Probability
0, 0	0	$0.9 \times 0.9 = 0.81$
0, 1	1	$0.9 \times 0.1 = 0.09$
1, 0	1	$0.1 \times 0.9 = 0.09$
1, 1	1	$0.1 \times 0.1 = 0.01$

Table 5.3. Possible sample values for a sample of size two.

The probability distribution for M is given in Table 5.4.

Value, m	0	1
$P(M = m)$	0.81	0.19

Table 5.4. Probability distribution for M.

Therefore $E(M) = 0 \times 0.81 + 1 \times 0.19 = 0.19$.

But the population maximum is 1 and so the sample maximum, which is Luc's estimator, underestimates the population maximum on average. Therefore the sample maximum is a biased estimator of the population maximum.

You have now seen that it is not always sensible to use the sample version of a statistic as an estimator of the population version of the same statistic. There is a particularly important case where this warning applies. The case in question is the estimation of population variance. Consider the following example.

Example 5.1.2

A jar contains a large number of small discs, 40% of which are numbered 1, 40% numbered 2 and 20% numbered 3. A random sample of size three is taken from the jar. By considering all possible values for samples of size three, determine whether the variance of the sample is an unbiased estimator of the variance of the population.

The mean of the population is

$$\mu = 1 \times 0.4 + 2 \times 0.4 + 3 \times 0.2 = 1.8.$$

The variance of the population is

$$\sigma^2 = 1^2 \times 0.4 + 2^2 \times 0.4 + 3^2 \times 0.2 - 1.8^2$$
$$= 0.56.$$

Consider the possible values of samples of size three. One such sample value is $1, 1, 2$.

The variance of this sample is $\frac{1}{3}\left(1^2 + 1^2 + 2^2\right) - \left(\frac{4}{3}\right)^2 = \frac{2}{9}$.

The chance of selecting this particular sample is $0.4 \times 0.4 \times 0.4 = 0.064$.

You can repeat this calculation for all 27 possible sample values. Table 5.5, below and on the next page, shows all possible sample values. It also shows the sample variance, V, and the probability of selecting that particular sample.

Sample values	Variance	Probability
1, 1, 1	0	$0.4^3 = 0.064$
2, 2, 2	0	$0.4^3 = 0.064$
3, 3, 3	0	$0.2^3 = 0.008$
1, 1, 2	$\frac{2}{9}$	$0.4^3 = 0.064$
2, 1, 1	$\frac{2}{9}$	$0.4^3 = 0.064$
1, 2, 1	$\frac{2}{9}$	$0.4^3 = 0.064$
1, 2, 2	$\frac{2}{9}$	$0.4^3 = 0.064$
2, 1, 2	$\frac{2}{9}$	$0.4^3 = 0.064$
2, 2, 1	$\frac{2}{9}$	$0.4^3 = 0.064$
1, 1, 3	$\frac{8}{9}$	$0.4^2 \times 0.2 = 0.032$
1, 3, 1	$\frac{8}{9}$	$0.4^2 \times 0.2 = 0.032$
3, 1, 1	$\frac{8}{9}$	$0.4^2 \times 0.2 = 0.032$
1, 3, 3	$\frac{8}{9}$	$0.4 \times 0.2^2 = 0.016$

Sample values	Variance	Probability
3, 1, 3	$\frac{8}{9}$	$0.4 \times 0.2^2 = 0.016$
3, 3, 1	$\frac{8}{9}$	$0.4 \times 0.2^2 = 0.016$
2, 2, 3	$\frac{2}{9}$	$0.4^2 \times 0.2 = 0.032$
2, 3, 2	$\frac{2}{9}$	$0.4^2 \times 0.2 = 0.032$
3, 2, 2	$\frac{2}{9}$	$0.4^2 \times 0.2 = 0.032$
2, 3, 3	$\frac{2}{9}$	$0.4 \times 0.2^2 = 0.016$
3, 2, 3	$\frac{2}{9}$	$0.4 \times 0.2^2 = 0.016$
3, 3, 2	$\frac{2}{9}$	$0.4 \times 0.2^2 = 0.016$
1, 2, 3	$\frac{6}{9}$	$0.4^2 \times 0.2 = 0.032$
1, 3, 2	$\frac{6}{9}$	$0.4^2 \times 0.2 = 0.032$
2, 1, 3	$\frac{6}{9}$	$0.4^2 \times 0.2 = 0.032$
2, 3, 1	$\frac{6}{9}$	$0.4^2 \times 0.2 = 0.032$
3, 1, 2	$\frac{6}{9}$	$0.4^2 \times 0.2 = 0.032$
3, 2, 1	$\frac{6}{9}$	$0.4^2 \times 0.2 = 0.032$

Table 5.5. Possible sample values and sample variance for a sample of size three.

From this table you can construct the probability distribution of V, as in Table 5.6.

Value, v	0	$\frac{2}{9}$	$\frac{6}{9}$	$\frac{8}{9}$
$P(V = v)$	0.136	0.528	0.192	0.144

Table 5.6. Probability distribution table for V.

Now $E(V) = 0 \times 0.136 + \frac{2}{9} \times 0.528 + \frac{6}{9} \times 0.192 + \frac{8}{9} \times 0.144 = 0.3733\ldots$.

Notice that $E(V)$ is not equal to 0.56. This means that V, the sample variance, is a biased estimator of the population variance.

It is possible, however, to modify the sample statistic V in order to obtain an unbiased estimator.

As a start, check that in the example above $E(V) = \frac{2}{3} \times 0.56$. This is an instance of a general rule.

If V is the sample variance of a sample of size n taken from a population with an unknown variance σ^2, then

$$E(V) = \frac{n-1}{n} \times \sigma^2.$$

Given this general rule and with a little thought about the meaning of expectation you should be able to see that $E\left(\dfrac{nV}{n-1}\right) = \sigma^2$.

This means that

$$V = \left(\frac{X_1^2 + X_2^2 + \ldots + X_n^2}{n}\right) - \overline{X}^2$$

is a biased estimator of σ^2, but that

$$\frac{nV}{n-1} = \frac{n}{n-1}\left(\left(\frac{X_1^2 + X_2^2 + \ldots + X_n^2}{n}\right) - \overline{X}^2\right)$$

is an unbiased estimator of σ^2.

The expression $\dfrac{n}{n-1}\left(\left(\dfrac{X_1^2 + X_2^2 + \ldots + X_n^2}{n}\right) - \overline{X}^2\right)$ is usually denoted by S^2, or by $\hat{\sigma}^2$.

The expression for S^2 can be written in Σ-notation as

$$S^2 = \frac{n}{n-1}\left(\frac{\sum X^2}{n} - \overline{X}^2\right) = \frac{1}{n-1}\left(\sum X^2 - n\overline{X}^2\right).$$

Substituting $\overline{X} = \dfrac{\sum X}{n}$ gives

$$S^2 = \frac{1}{n-1}\left(\sum X^2 - n\left(\frac{\sum X}{n}\right)^2\right) = \frac{1}{n-1}\left(\sum X^2 - \frac{(\sum X)^2}{n}\right).$$

Recall that in S1 Section 3.7 you saw that $\dfrac{\sum X^2}{n} - \overline{X}^2 = \dfrac{1}{n}\sum(X - \overline{X})^2$. Using this substitution in the previous equation you can define S^2 alternatively as

$$S^2 = \frac{n}{n-1}\left(\frac{1}{n}\sum(X - \overline{X})^2\right) = \frac{1}{n-1}\sum(X - \overline{X})^2.$$

A proof of this result is given in Section 5.2.

The following is a summary of the methods for obtaining unbiased estimates of the mean and variance of a population from sample values $x_1, x_2, \ldots, x_n$.

If you take a random sample with values $x_1, x_2, \ldots, x_n$ from a population,

- an unbiased estimate of the mean, μ, of the population is the sample mean,

- an unbiased estimate of the variance, σ^2, of the population is

$$s^2 = \frac{1}{n-1}\left(\sum x^2 - \frac{\left(\sum x\right)^2}{n}\right) = \frac{1}{n-1}\sum (x - \bar{x})^2.$$

Notice that the upper-case letters $X_1, X_2, \ldots, X_n$ and S have been replaced here by lower-case letters $x_1, x_2, \ldots, x_n$ and s because the definition refers to a set of sample values which you have actually obtained.

You now know two different formulae for variance: the formula from S1 Chapter 3,

$$\frac{\sum x^2}{n} - \bar{x}^2 \qquad \text{or its equivalent} \qquad \frac{1}{n}\sum (x - \bar{x})^2,$$

and the formula that was just derived,

$$s^2 = \frac{1}{n-1}\left(\sum x^2 - \frac{\left(\sum x\right)^2}{n}\right) = \frac{1}{n-1}\sum (x - \bar{x})^2.$$

You do not have a choice, however, about which formula to use in a given situation. The context of your calculation will determine which formula you must use.

If the values $x_1, x_2, \ldots, x_n$ represent the *whole population*, that is all the values of interest, and you wish to calculate the variance of that population, then use

$$\sigma^2 = \frac{\sum x^2}{n} - \bar{x}^2 \text{ or its equivalent } \frac{1}{n}\sum (x - \bar{x})^2.$$

If, on the other hand, you are trying to estimate the variance of a larger population from which the values $x_1, x_2, \ldots, x_n$ are a *sample*, then use

$$s^2 = \frac{1}{n-1}\left(\sum x^2 - \frac{\left(\sum x\right)^2}{n}\right) \text{ or its equivalent } \frac{1}{n-1}\sum (x - \bar{x})^2.$$

Most calculators have separate keys for these two different variances.

If the sample values are given as grouped data, an unbiased estimate of the population variance is given by

$$s^2 = \frac{1}{\left(\sum f\right)-1}\left(\sum x^2 f - \frac{\left(\sum xf\right)^2}{\sum f}\right).$$

Example 5.1.3

(a) Nine CDs were played and the playing time of each CD was recorded. The times, in minutes, are given below.

    49,   56,   55,   68,   61,   57,   61,   52,   63

Find the mean playing time of the nine CDs and the variance of the playing times.

(b) A student was doing a project on the playing times of CDs. She wished to estimate the mean playing time for CDs sold throughout the country and she wished also to estimate the variance of playing times of CDs sold throughout the country. She took a sample of nine CDs and recorded their playing times. The results are given below.

    49,   56,   55,   68,   61,   57,   61,   52,   63

(i) Use the student's data to estimate the mean playing time for CDs sold in the country.

(ii) Use the student's data to estimate the variance of the playing times of CDs sold in the country.

Notice that the two parts, (a) and (b) look very similar. However in part (a) you are only interested in the playing times of the nine CDs which have been selected. In part (b) you wish to make estimates of population parameters from the sample data that you have been given.

(a) $\bar{x} = \frac{1}{9}(49 + 56 + \ldots + 63) = 58$, so the mean is 58 minutes.

$$\text{Variance} = \frac{\sum x^2}{n} - \bar{x}^2 = \frac{1}{9}\left(49^2 + 56^2 + \ldots + 63^2\right) - 58^2$$
$$= 30.4\ldots,$$

so the variance is 30.4 min^2, correct to 3 significant figures.

(b) In this part you are estimating the mean and variance of the population.

(i) Recall that the unbiased estimator of the population mean, μ, is the mean of the sample $\bar{x}$, so the estimate of μ will be 58 minutes.

(ii) To obtain an unbiased estimate of σ^2, you need to use $s^2 = \frac{1}{n-1}\left(\sum x^2 - \frac{\left(\sum x\right)^2}{n}\right)$.

$$s^2 = \frac{1}{9-1}\left(49^2 + 56^2 + \ldots + 63^2\right) - \frac{(49 + 56 + \ldots + 63)^2}{9}$$
$$= 34.25 = 34.3, \text{ correct to 3 significant figures.}$$

In the next example you are not given individual data values for the sample. Instead the data are summarised in a grouped frequency table.

Example 5.1.4

A fishing crew recorded the masses in kilograms of 200 fish of a particular species that were caught on their trawler. The results are summarised in the table below. The weights given are mid-class values.

Weight of fish (kg)	0.5	1.25	1.75	2.25	2.75	3.5	4.5	5.5	7.0	10.5
Number of fish in class	21	32	33	24	18	21	16	12	11	12

Assuming that these fish are a random sample from the population of this species, estimate
(a) the mean mass, in kilograms, of a fish of this species,
(b) the variance of masses of fish of this species.

Weight of fish, x (kg) (mid-class value)	Number of fish in class, f	xf	$x^2 f$
0.5	21	10.5	5.25
1.25	32	40	50
1.75	33	57.75	101.0...
2.25	24	54	121.5
2.75	18	49.5	136.1...
3.5	21	73.5	257.2...
4.5	16	72	324
5.5	12	66	363
7	11	77	539
10.5	12	126	1323
Total	200	626.25	3220.1875

(a) To find the mean of the sample use the formula

$$\bar{x} = \frac{\sum xf}{\sum f} = \frac{626.25}{200} = 3.131\,25 \text{, so the sample mean is } 3.131\,25 \text{ kg.}$$

Since the sample mean is an unbiased estimator of the population mean, you can take this value as the estimate of the mean mass of all fish of this species.

Therefore the estimate of the mean mass of fish of this species is 3.131 kg.

(b) An unbiased estimate of the population variance is given by

$$s^2 = \frac{1}{(\sum f) - 1}\left(\sum x^2 f - \frac{(\sum xf)^2}{\sum f}\right)$$

$$= \frac{1}{200 - 1}\left(3220.1875 - \frac{626.25^2}{200}\right) = 6.327...$$

$$= 6.33 \text{, correct to 3 significant figures.}$$

5.2* Proof that $S^2 = \dfrac{\sum(X_i - \overline{X})^2}{n-1}$ **is an unbiased estimator of** σ^2

You may omit this section if you wish.

Let n independent observations be made of a random variable, X, with mean μ and variance σ^2.

$$S^2 = \frac{\sum(X_i - \overline{X})^2}{n-1} = \frac{n}{n-1}\left(\frac{\sum X_i^2}{n} - \overline{X}^2\right) = \frac{1}{n-1}\left(\sum X_i^2 - n\overline{X}^2\right).$$

Consider

$$E\left(\sum X_i^2 - n\overline{X}^2\right) = E\left(X_1^2\right) + E\left(X_2^2\right) + \ldots + E\left(X_n^2\right) - nE\left(\overline{X}^2\right)$$

$$= nE\left(X^2\right) - nE\left(\overline{X}^2\right). \tag{5.1}$$

Now $\mathrm{Var}(X) = \sigma^2 = E\left(X^2\right) - (E(X))^2 = E\left(X^2\right) - \mu^2$, so

$$E\left(X^2\right) = \sigma^2 + \mu^2. \tag{5.2}$$

As $\overline{X}$ has mean μ and variance $\dfrac{\sigma^2}{n}$,

$$\mathrm{Var}\left(\overline{X}\right) = E\left(\overline{X}^2\right) - \left(E\left(\overline{X}\right)\right)^2 \quad \Leftrightarrow \quad \frac{\sigma^2}{n} = E\left(\overline{X}^2\right) - \mu^2,$$

giving

$$E\left(\overline{X}^2\right) = \frac{\sigma^2}{n} + \mu^2. \tag{5.3}$$

Substituting from Equations 5.2 and 5.3 into Equation 5.1 gives

$$E\left(\sum X_i^2 - n\overline{X}^2\right) = n\left(\sigma^2 + \mu^2\right) - n\left(\frac{\sigma^2}{n} + \mu^2\right) = (n-1)\sigma^2.$$

So $E\left(S^2\right) = E\left(\dfrac{1}{n-1}\left(\sum X_i^2 - n\overline{X}^2\right)\right) = \sigma^2$ and $S^2 = \dfrac{\sum(X_i - \overline{X})^2}{n-1}$ is an unbiased estimator of σ^2.

Exercise 5A

1 A random sample of 10 people working for a certain company with 4000 employees are asked, at the end of a day, how much they had spent on lunch that day. The results in \$ are as follows.

| 1.98 | 1.84 | 1.75 | 1.94 | 1.56 | 1.88 | 1.05 | 2.10 | 1.85 | 2.35 |

Calculate unbiased estimates of the mean and variance of the amounts spent on lunch that day by all workers employed by the company.

2 The diameters of 20 randomly chosen plastic doorknobs of a certain make were measured. The results, x cm, are summarised by $\sum x = 102.3$ and $\sum x^2 = 523.54$. Find

(a) the variance of the diameters in the sample,

(b) an unbiased estimate of the variance of the diameters of all knobs produced.

3 The number of vehicle accidents occurring each day along a long stretch of a particular road was monitored for a period of 100 randomly chosen days. The results are summarised in the following table.

Number of accidents	0	1	2	3	4	5	6
Number of days	8	12	27	35	13	4	1

(a) Find unbiased estimates of the mean and variance of the daily number of accidents.

(b) Estimate the probability that the mean number of accidents per day over a period of a month of 30 days is greater than 2.7.

Give a reason why your estimate in part (b) may be considerably in error.

4 A random sample of 150 pebbles was collected from a beach. The masses of the pebbles, correct to the nearest gram, are summarised in the following grouped frequency table.

Mass (g)	10–19	20–29	30–39	40–49	50–59	60–69	70–79	80–89
Frequency	1	4	22	40	49	28	4	2

(a) Find, to 3 decimal places, unbiased estimates of the mean and variance of the masses of all the pebbles on the beach.

(b) Estimate the probability that the mean mass of a random sample of 40 pebbles from the same beach is less than 50 g. Give your answer correct to 2 decimal places.

5 Unbiased estimates of the mean and variance of a population, based on a random sample of 24 observations, are 5.5 and 2.42 respectively. Another random observation of 8.0 is obtained. Find new unbiased estimates of the mean and variance with this new information.

Assuming that the sample mean has a normal distribution, estimate the probability that a sample mean based on a sample size of 25 is within 0.01 of the population mean. Explain why your answer is only an estimate.

6 Thirty oranges are chosen at random from a large box of oranges. Their masses, x grams, are summarised by $\sum x = 3033$ and $\sum x^2 = 306\,676$. Find, to 4 significant figures, unbiased estimates for the mean and variance of the mass of an orange in the box.

The oranges are packed in bags of 10 in a shop and the shopkeeper told customers that most bags weigh more than a kilogram. Show that the shopkeeper's statement is correct indicating any necessary assumption made in your calculation.

5.3   The concept of a confidence interval

In Section 5.1 you learnt that the mean, $\overline{X}$, of a random sample is an unbiased estimator of the population mean, μ. For example, to estimate the mean amount of pocket money received by all the children in a primary school you could take a random sample of the children and ask each child how much pocket money he or she receives. Suppose you find that $\sum x = 111.50$ for a random sample of 50 children, where x is measured in dollars. Then an unbiased estimate of the population mean, μ, is given by

$$\overline{x} = \frac{\sum x}{n} = \frac{111.50}{50} = 2.23.$$

Such a value is sometimes called a **point estimate** because it gives an estimate of the population mean in the form of a single value or 'point' on a number line. Such values are useful, for example, in comparing populations. However, since $\overline{X}$ is a random variable, the value which it takes will vary from sample to sample. As a result you have no idea how close to the actual population mean a point estimate is likely to be. The purpose of a 'confidence interval' is to give an estimate in a form which also indicates the estimate's likely accuracy. A **confidence interval for the mean** is a range of values which has a given probability of 'trapping' the population mean. It is usually taken to be symmetrical about the sample mean. So, if the sample mean takes the value $\overline{x}$, the associated confidence interval would be $[\overline{x} - c, \overline{x} + c]$ where c is a number whose value has yet to be found.

Notice the notation for an interval which is being used. The interval $[\overline{x} - c, \overline{x} + c]$ means the real numbers from $\overline{x} - c$ to $\overline{x} + c$, including the end-points. For example, $[-2.1, 6.8]$ means real numbers y such that $-2.1 \leqslant y \leqslant 6.8$.

In Fig. 5.7 the sample mean $\overline{X}$ takes a value $\overline{x}_1$ and the confidence interval covers the range of values $[\overline{x}_1 - c, \overline{x}_1 + c]$. In this case the confidence interval traps the population mean, μ. For a different sample, with a different sample mean, $\overline{x}_2$, the confidence interval might not trap μ. This situation is illustrated in Fig. 5.8.

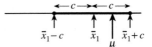

Fig. 5.7. Confidence interval which traps the population mean μ.

The end-points of the confidence interval are themselves random variables since they vary from sample to sample. They can be written as $\overline{X} - c$ and $\overline{X} + c$. You can see from Fig. 5.7 and Fig. 5.8 that the confidence interval will trap μ if the difference between the sample mean and the population mean is less than or equal to c. Expressed algebraically this condition is $|\overline{X} - \mu| \leqslant c$.

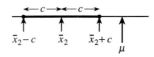

Fig. 5.8. Confidence interval which does not trap the population mean μ.

The next section explains how the value of c is chosen so as to give a specified probability that the confidence interval traps μ.

5.4   Calculating a confidence interval

Consider the following situation. The masses of tablets produced by a machine are known to be distributed normally with a standard deviation of 0.012 g. The mean mass of the tablets produced is monitored at regular intervals by taking a sample of 25 tablets and calculating the sample mean, $\overline{X}$.

Suppose you wish to find an interval which has a 95% probability of trapping the population mean, μ. The mass, X (in grams), of a single tablet is distributed normally with unknown mean μ and standard deviation 0.012; that is, $X \sim N(\mu, 0.012^2)$. So, for a sample of size 25, $\overline{X} \sim N\left(\mu, \dfrac{0.012^2}{25}\right)$ (see Section 4.6).

Using the notation of Section 5.3, the population mean is trapped in the interval $\left[\overline{X} - c, \overline{X} + c\right]$, where c is a constant, if $\left|\overline{X} - \mu\right| \leqslant c$. The interval has a probability of 95% of trapping the population mean if, and only if, $P\left(\left|\overline{X} - \mu\right| \leqslant c\right) = 0.95$.

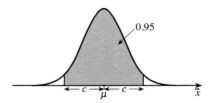

Fig. 5.9 Diagram to illustrate $P\left(\left|\overline{X} - \mu\right| \leqslant c\right) = 0.95$.

Fig. 5.9 shows the sampling distribution of $\overline{X}$ with this probability indicated. The value of c is found by standardising $\left|\overline{X} - \mu\right|$ to give $Z = \dfrac{\overline{X} - \mu}{\sqrt{\dfrac{0.012^2}{25}}}$ where $Z \sim N(0,1)$.

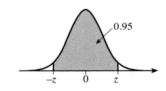

Fig. 5.10. Standardisation of Fig. 5.9.

Fig. 5.10 shows the distribution of Z with the probability of 0.95 indicated. The unshaded areas each have probability $\frac{1}{2} \times 0.05 = 0.025$ (see also Fig. 5.13 on page 90). Therefore the value of z in this diagram is given by $\Phi(z) = 0.975$. This could be found using the normal distribution tables 'in reverse', but it is convenient to use the 'critical values' table at the bottom of page 165. Taking $p = 0.975$ gives $z = 1.96$.

Thus $1.96 = \dfrac{\overline{X} - \mu}{\sqrt{\dfrac{0.012^2}{25}}} = \dfrac{\overline{X} - \mu}{\dfrac{0.012}{\sqrt{25}}}$. But $\overline{X} - \mu = c$, so

$$1.96 = \frac{c}{\dfrac{0.012}{\sqrt{25}}} \qquad (5.4)$$

giving $c = 1.96 \dfrac{0.012}{\sqrt{25}} = 0.004\ 70$ to 3 significant figures.

So the interval which has a 95% probability of trapping μ is $\left[\overline{X} - 0.0047, \overline{X} + 0.0047\right]$.

Suppose that you took a sample of 25 tablets and found that the mean mass was, for example, 0.5642 g. This sample mean would give a value for the interval of $[0.5642 - 0.0047, 0.5642 + 0.0047]$, which is $[0.5595, 0.5689]$. Such an interval is called a **95% confidence interval for the population mean**.

It is important to realise that such a confidence interval may or may not trap μ, depending on the value of $\overline{X}$. Suppose, for a moment, that you know the value of μ and you take a number of different samples of 25 tablets. Fig. 5.11 shows confidence intervals calculated in the way described above for 30 different samples of 25 tablets.

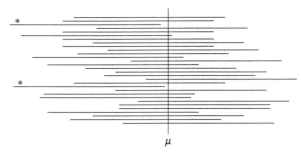

The majority of the confidence intervals trap μ but a few (marked *) do not. On average, the proportion of 95% confidence intervals which trap μ is 95%. In practice, of course, you do not know μ and would usually take only one sample and from it calculate one confidence interval. In this situation, you cannot *know* whether your particular confidence

Fig. 5.11. 95% confidence intervals calculated from 30 random samples.

interval does include μ: all you can say is that, on average, 95 times out of 100 it will contain μ.

The method which has been described for finding a 95% confidence interval for the mean can be generalised to a sample of size n from a normal population with standard deviation σ. Replacing 25 by n and 0.012 by σ in Equation 5.4 gives

$$1.96 = \frac{c}{\dfrac{\sigma}{\sqrt{n}}},$$

which, on rearranging, gives $c = 1.96 \dfrac{\sigma}{\sqrt{n}}$.

Given a sample of size n from a normal population with variance σ^2, a 95% confidence interval for the population mean is given by

$$\left[\overline{x} - 1.96 \frac{\sigma}{\sqrt{n}}, \ \overline{x} + 1.96 \frac{\sigma}{\sqrt{n}} \right], \qquad (5.5)$$

where $\overline{x}$ is the sample mean.

Computer activity

This activity requires access to a computer.

Make a spreadsheet containing 500 samples of size 25 from a population which is distributed as $N(0.56, 0.012^2)$. Using the facilities of the spreadsheet, calculate the 95% confidence interval for the mean for each sample. How many of these confidence intervals trap the population mean, 0.56? (Your answer should be reasonably close to 475. Why?) Explain why your answer need not necessarily be 475 exactly.

Example 5.4.1

The lengths of nails produced by a machine are known to be distributed normally with mean μ mm and standard deviation 0.7 mm. The lengths, in mm, of a random sample of 5 nails are 107.29, 106.56, 105.94, 106.99, 106.47.

(a) Calculate a symmetric 95% confidence interval for μ, giving the end-points correct to 1 decimal place.

(b) Two hundred random samples of 5 nails are taken and a symmetric 95% confidence interval for μ is calculated for each sample. Find the expected number of intervals which do not contain μ.

(a) The mean of the sample is given by

$$\bar{x} = \frac{107.29 + 106.56 + 105.94 + 106.99 + 106.47}{5} = 106.65.$$

Substituting this value of $\bar{x}$ together with $\sigma = 0.7$ and $n = 5$ into Equation 5.5 gives a symmetric 95% confidence interval for μ of

$$\left[106.65 - 1.96 \frac{0.7}{\sqrt{5}}, \ \ 106.65 + 1.96 \frac{0.7}{\sqrt{5}} \right],$$

which, on simplifying, gives [106.036...,107.263...]. So, to 1 decimal place, the symmetric 95% confidence interval for μ, measured in mm, is [106.0,107.3].

(b) On average, 95% of the confidence intervals should include μ. This means that 5% will not include μ. So out of 200 confidence intervals you would expect $200 \times 5\% = 10$ not to include μ.

Example 5.4.2

For a method of measuring the velocity of sound in air, the results of repeated experiments are known to be distributed normally with standard deviation 6 m s^{-1}. A number of measurements are made using this method, and from these measurements a symmetric 95% confidence interval for the velocity of sound in air is calculated. Find the width of this confidence interval for   (a) 4,   (b) 36 measurements.

A symmetric 95% confidence interval extends from $\bar{x} - 1.96 \dfrac{\sigma}{\sqrt{n}}$ to $\bar{x} + 1.96 \dfrac{\sigma}{\sqrt{n}}$,

so its width is $\bar{x} + 1.96 \dfrac{\sigma}{\sqrt{n}} - \left(\bar{x} - 1.96 \dfrac{\sigma}{\sqrt{n}} \right) = 2 \times 1.96 \dfrac{\sigma}{\sqrt{n}}$.

In this example $\sigma = 6$, so the width of the confidence interval is $2 \times 1.96 \dfrac{6}{\sqrt{n}} = \dfrac{23.52}{\sqrt{n}}$.

(a) For $n = 4$, the width of the confidence interval is $\dfrac{23.52}{\sqrt{4}} = 11.76$.

(b) For $n = 36$, the width of the confidence interval is $\dfrac{23.52}{\sqrt{36}} = 3.92$.

Note that nine times as many measurements are required to reduce the width of the confidence interval by a factor of 3.

5.5 Different levels of confidence

As Example 5.4.1 makes clear, there is a probability of 5 in 100 that a 95% confidence interval does not include μ. There are circumstances in which you may wish to be more certain that the confidence interval which you have calculated does include μ. For example, you may wish to be 99% certain. Fig. 5.12 is the diagram corresponding to Fig. 5.9 for this situation. The only difference in the calculation of the confidence interval is that a different value of z is needed. In this case $\Phi(z) = 0.995$, giving $z = 2.576$, so that the 99% confidence interval for the mean is given by

$$\left[\bar{x} - 2.576 \frac{\sigma}{\sqrt{n}}, \ \bar{x} + 2.576 \frac{\sigma}{\sqrt{n}} \right].$$

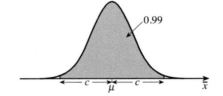

Note that this 99% confidence interval is wider than the 95% confidence interval: this is to be expected since it is more likely to trap μ than the 95% interval. If you want a 90% confidence interval, then the appropriate value of z is that value for which $\Phi(z) = 0.95$ giving $z = 1.645$ and the confidence interval is narrower. You can see that there is a balance between precision and certainty: if you

Fig. 5.12. Diagram to illustrate $P(|\bar{X} - \mu| \le c) = 0.99$.

increase one you decrease the other. In order to increase both the precision and the certainty of a confidence interval you have to increase the sample size.

In general, a $100(1 - \alpha)\%$ confidence interval for the population mean for a sample of size n taken from a normal population with variance σ^2 is given by

$$\left[\bar{x} - z \frac{\sigma}{\sqrt{n}}, \bar{x} + z \frac{\sigma}{\sqrt{n}} \right], \qquad (5.6)$$

where $\bar{x}$ is the sample mean and the value of z is such that $\Phi(z) = 1 - \frac{1}{2}\alpha$. The relationship between α and z is illustrated in Fig. 5.13.

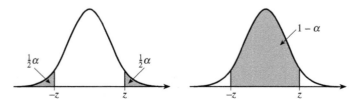

Fig. 5.13. Diagram to illustrate the relationship between critical values and confidence intervals.

Example 5.5.1

The measurement error made in measuring the concentration in parts per million (ppm) of nitrate ions in water by a particular method is known to be distributed normally with mean 0 and standard deviation 0.05.

(a) If 10 measurements on a specimen gave $\sum x = 11.37$ ppm, determine a symmetric 99.5% confidence interval for the true concentration, μ, of nitrate ions in the specimen, giving the end-points of the interval correct to 2 decimal places.

(b) How many measurements would be required in order to reduce the width of this interval to 0.03 ppm at most?

(a) The measured value, X, of the nitrate ion concentration is equal to $\mu + Y$ where Y is the measurement error. Thus

$$E(X) = E(\mu + Y) = \mu + E(Y) = \mu + 0 = \mu \qquad \text{(using Equation 2.1)}$$

and $\quad \mathrm{Var}(X) = \mathrm{Var}(\mu + Y) = \mathrm{Var}(Y) \qquad \text{(using Equation 2.2)}$

$$= 0.05^2.$$

This means that $X \sim N(\mu, 0.05^2)$. For a 99.5% confidence interval, $1 - \alpha = 0.995$, so $\alpha = 0.005$ and $\frac{1}{2}\alpha = 0.0025$. Thus $\Phi(z) = 1 - \frac{1}{2}\alpha = 1 - 0.0025 = 0.9975$. From the table at the bottom of page 165, taking $p = 0.9975$ gives $z = 2.807$. So a 99.5% confidence interval for μ is

$$\left[\bar{x} - 2.807\frac{\sigma}{\sqrt{n}}, \ \bar{x} + 2.807\frac{\sigma}{\sqrt{n}}\right] = \left[\frac{11.37}{10} - 2.807\frac{0.05}{\sqrt{10}}, \ \frac{11.37}{10} + 2.807\frac{0.05}{\sqrt{10}}\right]$$

$$= [1.092\ldots, 1.181\ldots]$$

$$= [1.09, 1.18], \quad \text{correct to 2 decimal places.}$$

(b) The 99.5% confidence interval for a sample of size n has width $2 \times 2.807\dfrac{0.05}{\sqrt{n}}$.

For this width to be at most 0.03, $2 \times 2.807\dfrac{0.05}{\sqrt{n}} \leqslant 0.03$.

Rearranging gives

$$\sqrt{n} \geqslant \frac{2 \times 2.807 \times 0.05}{0.03}, \quad \text{so} \quad n \geqslant 87.5\ldots.$$

Since n must be an integer, 88 or more measurements are required to give a confidence interval of width 0.03 ppm at most.

Exercise 5B

1 Bags of sugar have masses which are distributed normally with mean μ grams and standard deviation 4.6 grams. The sugar in each of a random sample of 5 bags taken from a production line is weighed, with the following results, in grams.

498.2 501.3 503.7 496.8 502.5

Calculate a symmetric 95% confidence interval for μ.

If a 95% symmetric confidence interval for μ was calculated for each of 200 samples of 5 bags, how many of the confidence intervals would be expected to contain μ?

2 The volume of milk in litre cartons filled by a machine has a normal distribution with mean μ litres and standard deviation 0.05 litres. A random sample of 25 cartons was selected and the contents, x litres, measured. The results are summarised by $\sum x = 25.11$. Calculate

(a) a symmetric 98% confidence interval for μ,

(b) the width of a symmetric 90% confidence interval for μ based on the volume of milk in a random sample of 50 cartons.

3 The random variable X has a normal distribution with mean μ and variance σ^2. A symmetric 90% confidence interval for μ based on a random sample of 16 observations of X has width 4.24. Find

(a) the value of σ,

(b) the width of a symmetric 90% confidence interval for μ based on a random sample of 4 observations of X,

(c) the width of a symmetric 95% confidence interval for μ based on a random sample of 4 observations of X.

4 In a particular country the heights of fully-grown males may be modelled by a normal distribution with mean 178 cm and standard deviation 7.5 cm. The 11 male (fully-grown) biology students present at a university seminar had a mean height of 175.2 cm. Assuming a standard deviation of 7.5 cm, and stating any further assumption, calculate a symmetric 99% confidence interval for the mean height of all fully-grown male biology students.

Does the confidence interval suggest that fully-grown male biology students have a different mean height from 178 cm?

5 A machine is designed to produce metal rods of length 5 cm. In fact, the lengths are distributed normally with mean 5.00 cm and standard deviation 0.032 cm. The machine is moved to a new site and, in order to check whether or not the mean length has altered, the lengths of a random sample of 8 rods are measured. The results, in cm, are as follows.

5.07 4.95 4.98 5.06 5.13 5.05 4.98 5.06

(a) Assuming that the standard deviation is unchanged, calculate a symmetric 95% confidence interval for the mean length of the rods produced by the machine in its new position.

(b) State, giving a reason, whether you consider that the mean length has changed.

6 A method used to determine the percentage of nitrogen in a fertiliser has an error which is distributed normally with zero mean and standard deviation 0.34%. Ten independent determinations of the percentage of nitrogen gave a mean value of 15.92%.

(a) Calculate a symmetric 98% confidence interval for the percentage of nitrogen in the fertiliser.

(b) Find the smallest number of extra independent determinations that would reduce the width of the symmetric 98% confidence interval to at most 0.4%.

5.6 Confidence interval for a large sample

In the examples considered so far it has been assumed that the samples are drawn from a normal distribution and that the variance of this distribution is known. In practice you will not always be sure that the population is normal and you may or may not have accurate information about its variance. However, the method of calculating a confidence interval which has been described in this chapter can still be applied provided that the sample is large.

Consider first the distribution from which the sample is taken. Although this may not be normal, the central limit theorem (Section 4.4) tells you that the sample mean, $\bar{X}$, is distributed approximately normally provided that the sample is large.

Next, consider the population variance. An unbiased estimate of this can be calculated from the sample using the equation on page 81.

$$s^2 = \frac{1}{n-1}\sum(x - \bar{x})^2 = \frac{1}{n-1}\left(\sum x^2 - \frac{(\sum x)^2}{n}\right).$$

This estimate will vary from sample to sample, but for large samples the variation is so small that s can be treated as though it is constant and can be used in place of σ in Equation 5.6. A rule of thumb is that 'large' means a sample size of 30 or more.

Given a *large* sample $(n \geqslant 30)$ from any population, a $100(1-\alpha)\%$ confidence interval for the population mean is given by

$$\left[\bar{x} - z\frac{s}{\sqrt{n}},\ \bar{x} + z\frac{s}{\sqrt{n}}\right] \qquad (5.7)$$

where $\bar{x}$ is the sample mean, the value of z is such that $\Phi(z) = 1 - \frac{1}{2}\alpha$ and

$$s^2 = \frac{1}{n-1}\left(\sum x^2 - \frac{(\sum x)^2}{n}\right).$$

Example 5.6.1

On 1 January 100 new Eternity light bulbs were installed in a certain building, together with a device which records how long each light bulb is used. By 1 March all 100 bulbs had failed. The data for the recorded lifetimes, t (in hours of use), are summarised by $\sum t = 10\,500$, $\sum t^2 = 1\,712\,500$. Assuming that the bulbs constituted a random sample of Eternity light bulbs, obtain a symmetric 99% confidence interval for the mean lifetime of Eternity light bulbs, giving your answer correct to the nearest hour. (OCR, adapted)

In this example neither the distribution of the population of lifetimes nor the population variance are known. However, the sample size is much larger than 30, so Equation 5.7 can be used. First it is necessary to calculate $\bar{t}$ and s^2.

$$\bar{t} = \frac{\sum t}{n} = \frac{10\,500}{100} = 105.$$

$$s^2 = \frac{1}{n-1}\left(\sum t^2 - \frac{\left(\sum t\right)^2}{n}\right) = \frac{1}{100-1}\left[1\,712\,500 - \frac{10\,500^2}{100}\right]$$

$$= 6161.6\ldots.$$

Thus $s = 78.49\ldots.$

For a 99% confidence interval, $1-\alpha = 0.99$, so $\alpha = 0.01$ and $\frac{1}{2}\alpha = 0.005$.

Thus $\Phi(z) = 1 - \frac{1}{2}\alpha = 1 - 0.005 = 0.995$.

From the table at the bottom of page 165, taking $p = 0.995$ gives $z = 2.576$.

So, substituting into Equation 5.7 gives a 99% confidence of the population mean as

$$\left[105 - 2.576\frac{78.49\ldots}{\sqrt{100}}, \ 105 + 2.576\frac{78.49\ldots}{100}\right]$$

$$= [85,125], \text{ to the nearest hour.}$$

Exercise 5C

1   During one year, a squash player bought 72 squash balls of a certain brand and recorded for each the time, x hours, before failing. The results are summarised by $\sum x = 372.4$ and $\sum x^2 = 2301.32$.

(a) Find a symmetric 90% confidence interval for the mean life of that brand of squash ball.

(b) State, and justify, any approximation used in your calculation.

2   The contents of 140 bags of flour selected randomly from a large batch delivered to a store are weighed and the results, w grams, summarised by $\sum(w-500) = -266$ and $\sum(w-500)^2 = 1178$.

(a) Calculate unbiased estimates of the batch mean and variance of the mass of flour in a bag.

(b) Calculate a symmetric 95% confidence interval for the batch mean mass.

The manager of the store believes that the confidence interval indicates a mean less than 500 g and considers the batch to be sub-standard. She has all of the bags in the batch weighed and finds that the batch mean mass is 501.1 g. How can this be reconciled with the confidence interval calculated in part (b)?

3   The depth of water in a lake was measured at 50 randomly chosen points on a particular day. The depths, d metres, are summarised by $\sum d = 366.45$ and $\sum d^2 = 2978.16$.

(a) Calculate an unbiased estimate of the variance of the depth of the lake.

(b) Calculate a symmetric 99% confidence interval for the mean depth of the lake.

The three months after the depths were obtained were very hot and dry and the water level in the lake dropped by 1.34 m. Based on readings taken at the same 50 points as before, which were all greater than 1.34 m, what will be a symmetric 99% confidence interval for the new mean depth?

4 An environmental science student carried out a study of the incidence of lichens on a stone wall. She selected, at random, 100 one-metre lengths of wall, all of the same height. The number of lichens in each section was counted and the results are summarised in the following frequency table.

Number of lichens	0	1	2	3	4	5	6
Number of sections	8	22	27	19	13	8	3

(a) Calculate the sample mean and an unbiased estimate of the population variance of the number of lichens per metre length of the wall.

(b) Calculate a symmetric 90% confidence interval for the mean number of lichens per metre length of the wall.

5 The pulse rates of 90 eight-year-old children chosen randomly from different schools in a city gave a sample mean of $\bar{x} = 86.6$ beats per minute and an unbiased estimate of population variance of $s^2 = 106.09$.

(a) Calculate a symmetric 98% confidence interval for the mean pulse rate of all eight-year-old children in the city.

(b) If, in fact, the 90 children were chosen randomly from those attending a hospital clinic, comment on how this information affects the interpretation of the confidence interval found in part (a).

6 Audio cassette tapes of a particular brand are claimed by the manufacturer to give, on average, at least 60 minutes of playing time. After receiving some complaints, the manufacturer's quality control manager obtains a random sample of 64 tapes and measures the playing time, t minutes, of each. The results are summarised by $\sum t = 3953.28$ and $\sum t^2 = 244\,557.00$.

(a) Calculate a symmetric 99% confidence interval for the population mean playing time of this brand of tape.

(b) Does the confidence interval support the customers' complaints? Give a reason for your answer.

5.7 Confidence interval for a proportion

Many statistical investigations are concerned with finding the proportion of a population which has a specified attribute. Suppose you were a manufacturer of a version of an appliance designed for left-handed people. In order to assess the potential market you would be interested in the proportion of left-handed people in the population. It would be impossible to ask everybody and so you would have to rely on a sample. Suppose that you were able to obtain information from a random sample of 500 people and you found that 60 of them were left-handed. It would seem reasonable to estimate that the proportion, p, of the population who are left-handed is $\frac{60}{500} = 0.12$, or 12%. However, you need to be certain that this method gives you an unbiased estimate of p.

Consider the more general situation where a random sample of n people are questioned. Provided that n is much smaller than the population size the distribution of X, the number of people in a sample of size n who are left-handed, will be $B(n, p)$ since

- there are a fixed number of trials (n people asked)
- each trial has two possible outcomes (left-handed or right-handed)
- the outcomes are mutually exclusive (assuming that no one is ambidextrous)
- the probability of a person being left-handed is constant
- the trials are independent.

Let P_s be the random variable 'the proportion of people in the sample who are left-handed'. Then $P_s = \dfrac{X}{n}$. The expected value of P_s is

$$\text{E}(P_s) = \text{E}\left(\frac{X}{n}\right) = \frac{1}{n}\text{E}(X) \qquad \text{(using Equation 2.3)}$$

$$= \frac{1}{n} \times np = p. \qquad \text{(since the mean of a binomial distribution is } np\text{)}$$

Thus the proportion in the sample does provide an unbiased point estimate of the population proportion.

It would be more useful, however, to find a confidence interval for p since this gives an idea of the precision of the estimate. In order to do this you need to consider the distribution of P_s. For a large sample, X will be distributed approximately normally and so P_s will also be distributed approximately normally (see Section 2.4). In order to calculate a confidence interval you also need the variance of P_s. This can be found as follows.

$$\text{Var}(P_s) = \text{Var}\left(\frac{X}{n}\right) = \frac{1}{n^2} \times \text{Var}(X) \qquad \text{(using Equation 2.4)}$$

$$= \frac{1}{n^2} \times npq = \frac{pq}{n} \qquad \text{(since the variance of a binomial distribution is } npq\text{)}.$$

If $X \sim B(n, p)$, then the sample proportion P_s, where $P_s = \dfrac{X}{n}$, is

distributed approximately as $N\left(p, \dfrac{pq}{n}\right)$ when n is large enough

to give $np > 5$ and $nq > 5$.

In practice, this will be achieved if there are more than 5 successes and 5 failures in n trials.

In Section 5.5 you saw that for the sampling distribution $\overline{X} \sim N\left(\mu, \dfrac{\sigma^2}{n}\right)$, a confidence interval for μ is given by $\left[\overline{x} - z\dfrac{\sigma}{\sqrt{n}}, \overline{x} + z\dfrac{\sigma}{\sqrt{n}}\right]$. By analogy a confidence interval for p

is found by replacing $\bar{x}$ by p_s, where $p_s = \dfrac{x}{n}$, and $\dfrac{\sigma}{\sqrt{n}}$ by $\sqrt{\dfrac{pq}{n}}$ to give an approximate

confidence interval of $\left[p_s - z\sqrt{\dfrac{pq}{n}},\ p_s + z\sqrt{\dfrac{pq}{n}} \right]$, where values of z are obtained as before.

This confidence interval is expressed in terms of p and q, which are not known (otherwise a confidence interval would not be required!). However, again similar to finding a confidence interval for the mean, these unknown quantities can be replaced by their estimates from the sample provided that the sample is large. The result is the interval

$$\left[p_s - z\sqrt{\dfrac{p_s q_s}{n}},\ p_s + z\sqrt{\dfrac{p_s q_s}{n}} \right], \qquad \text{where } q_s = 1 - p_s = 1 - \dfrac{x}{n}.$$

Strictly speaking, since a discrete distribution has been replaced by a continuous one in this calculation, a continuity correction should be applied. However, this is usually ignored when calculating a confidence interval for a proportion and it will be ignored here. Also, strictly speaking, an unbiased estimate of the population variance, given by $\dfrac{n}{n-1}\left(\dfrac{p_s q_s}{n} \right)$ rather than $\dfrac{p_s q_s}{n}$, should be used. The term $\dfrac{n}{n-1}$ is an adjustment similar to that in the formula for s^2. When n is large this adjustment makes little difference, and it will be ignored here.

Given a large random sample, size n, from a population in which a proportion of members, p, has a particular attribute, an approximate $100(1 - \alpha)\%$ confidence interval for p is

$$\left[p_s - z\sqrt{\dfrac{p_s q_s}{n}},\ p_s + z\sqrt{\dfrac{p_s q_s}{n}} \right], \qquad (5.8)$$

where p_s is the sample proportion with this attribute, $q_s = 1 - p_s$ and the value of z is such that $\Phi(z) = 1 - \tfrac{1}{2}\alpha$.

This confidence interval is approximate because

- a discrete distribution has been approximated by a continuous one
- a continuity correction has not been applied
- the population variance is estimated from the sample and the estimate used is biased
- the distribution of P_s is only approximately normal.

Equation 5.8 can now be used to calculate a confidence interval for the proportion of left-handers in the population. For the sample taken, $n = 500$, $p_s = 0.12$ and $q_s = 0.88$, and for a 95% confidence interval, $z = 1.96$. This gives a 95% confidence interval

$$\left[0.12 - 1.96\sqrt{\frac{0.12 \times 0.88}{500}}, \ 0.12 + 1.96\sqrt{\frac{0.12 \times 0.88}{500}}\right] = [0.092, 0.148]$$

where the answers have been given to 3 decimal places.

The same answer is obtained if a continuity correction is made and an unbiased estimate of variance is used.

It is interesting to note that the answer does not depend on the size of the population, only the size of the sample. This will always be true if the sample size is much less than the population size so that the value of p is effectively constant. This fact can be used to calculate the sample size required to give a confidence interval of specified width as illustrated in the following example.

Example 5.7.1

An opinion poll is to be carried out to estimate the proportion of the electorate of a country who will vote 'yes' in a forthcoming referendum. In a trial run a random sample of 100 people were questioned; 42 said they would vote 'yes'. Estimate the random sample size required to give a 99% confidence interval of the proportion with a width of 0.02.

For a 99% confidence interval, z takes the value 2.576 in Equation 5.8, so the width of this interval is $2 \times 2.576\sqrt{\dfrac{p_s q_s}{n}}$, where p_s and q_s are used to estimate p and q. The trial run gives $p_s = 0.42$ and hence $q_s = 0.58$. Thus the required value of n is

$$2 \times 2.576\sqrt{\frac{0.42 \times 0.58}{n}} = 0.02.$$

Rearranging and solving for n gives 16 165 to the nearest integer. This answer will only be approximate, for all four of the reasons in the box on page 97.

It is interesting to note that, if this example referred to a certain country with an electorate of about 40 million, then the sample required is only about 0.04% of the electorate. The problem lies in obtaining a random sample. In practice, opinion polls do not rely on random samples but use sophisticated techniques which are meant to ensure representative samples.

Exercise 5D

1 For the following sample sizes, n, and sample proportions, p_s, determine whether Equation 5.8 may be applied to find a valid confidence interval for the population proportion, p. Check whether the sample size is large enough.

    (a) $n = 60, p_s = 0.1$ (b) $n = 100, p_s = 0.04$ (c) $n = 14, p_s = 0.5$

2 For those cases in Question 1 where Equation 5.8 is suitable, calculate symmetric 95% confidence intervals for p.

3 In a study of computer usage a random sample of 200 private households in a particular town was selected and the number that own at least one computer was found to be 68. Calculate a symmetric 90% confidence interval for the percentage of households in the town that own at least one computer.

4 Of 500 cars passing under a bridge on a busy road 92 were found to be red.

  (a) Find a symmetric 98% confidence interval of the population proportion of red cars.

  (b) State any assumption required for the validity of the interval.

  (c) Describe a suitable population to which the interval applies.

  (d) If this experiment is carried out 50 times, what is the expected number of confidence intervals which would contain the population proportion of red cars?

5 A biased dice was thrown 600 times and 224 sixes were obtained. Calculate a symmetric 99% confidence interval of p, the probability of obtaining a six in a single throw of the dice.

Estimate the smallest number of times the dice should be thrown for the width of the symmetric 99% confidence interval of p to be at most 0.08.

Give two reasons why the previous answer is only an estimate.

6 Two machines, machine 1 and machine 2, produce shirt buttons. Random samples of 80 buttons are selected from the output of each machine and each button is inspected for faults. The sample from machine 1 contained 8 faulty buttons and that from machine 2 contained 19 faulty buttons.

  (a) Calculate symmetric 90% confidence intervals for the proportions, p_1 and p_2, of faulty buttons produced by each machine.

  (b) State, giving a reason, what the confidence intervals indicate about the relative sizes of p_1 and p_2.

5.8 Practical activities

1 Simulating an opinion poll   The 'population' for this activity consists of the outcomes of throwing a dice. A throw of 1 or 2 represents a person who would answer 'yes' to a particular question in an opinion poll while a throw of 3, 4, 5 or 6 represents a person who would answer 'no'. Each member of the class takes a sample of size 30 by throwing a dice 30 times, notes the number of 'people' who would vote 'yes' and then calculates a 95% confidence interval for the proportion of people in the population who would vote 'yes'. The confidence intervals for the class should be displayed on a diagram like that in Fig. 5.11. This diagram should also show the exact value for the proportion of people in the population who would vote 'yes', which in this case is $\frac{1}{3}$. Comment on the results.

This activity can be repeated with different values for the proportion who would vote 'yes', with different confidence intervals and with different sizes of samples.

2 How long is a 'foot'?   Measure the foot length in cm of at least 30 adult male subjects in (a) socks and (b) shoes. Use your results to calculate a confidence interval for the population mean for each set of measurements. Compare your results with those obtained by other members of the class. Do any of the students' confidence intervals contain the value for the imperial unit of length of 1 foot which is equal to 30.48 cm? Discuss: to what extent was your sample a random one from a well-defined population?

Miscellaneous exercise 5

1   The mean of a random sample of n observations of the random variable X, where
    $X \sim N(\mu, \sigma^2)$, is denoted by $\overline{X}$. In the following cases, find the value of $b - a$ in terms of
    n and σ.

    (a) $P(\overline{X} < a) = 0.03$, $P(\overline{X} > b) = 0.07$

    (b) $P(\overline{X} < a) = 0.04$, $P(\overline{X} > b) = 0.06$

    (c) $P(\overline{X} < a) = 0.05$, $P(\overline{X} > b) = 0.05$

    What do these results confirm about the width of a symmetric confidence interval for μ?

2   The masses of chocolate bars coming off a production line are normally distributed with
    mean μ and standard deviation 1.5 grams. The masses, in grams, of 9 randomly chosen
    bars are

        251.79   251.13   251.34   248.66   249.74   249.91   250.45   249.32   251.36.

    Calculate a 95% symmetric confidence interval for the population mean, μ.

3   The lengths of times, x hours, for which cars were parked in a town centre carpark were
    measured for a random sample of 200 cars, and it was found that $\sum x = 358.2$ and
    $\sum x^2 = 773.18$. Denoting the population mean and variance of the parking times by μ
    and σ^2 respectively, calculate

    (a) unbiased estimates of μ and σ^2,   (b) a symmetric 90% confidence interval for μ.

    It was also found that 39 of the 200 cars stayed in the carpark for more than $2\frac{1}{2}$ hours.
    Calculate a symmetric 90% confidence interval for the population proportion of cars
    staying for longer than $2\frac{1}{2}$ hours, giving the end-points to 3 decimal places.

    The manager of the carpark believes that only 10% of all cars stay for longer than
    $2\frac{1}{2}$ hours. Is this supported by your calculations?   (OCR, adapted)

4   The volume of paint dispensed into litre cans was measured for 100 randomly chosen cans
    and the results, x litres, are summarised by $\sum x = 104.0$ and $\sum x^2 = 110.06$.

    (a) Calculate unbiased estimates of the mean and variance of the volumes dispensed into
         the cans.

    (b) Calculate a symmetric 90% confidence interval for the mean volume, giving the end-
         points to a suitable degree of accuracy.

    (c) Estimate the smallest sample size required to give a symmetric 90% confidence
         interval of width at most 0.02 litres.   (OCR, adapted)

5   The board of trustees of a charitable trust wishes to make a change to the trust's
    constitution. In order for this to happen at least two-thirds of the members must vote for the
    change. Before the vote is taken, the secretary consults a random sample of 60 members
    and finds that 75% of them will vote for the change. Calculate a symmetric 95%
    confidence interval for the proportion of all members who will vote for the change.

    Nearer the time at which the vote is to be taken, the secretary consults a random sample of
    n members and finds, again, that 75% of them will vote for the change. Using this figure,
    he calculates a symmetric 99% confidence interval for the proportion of members who will
    vote for the change. This interval does not include the value two-thirds. Find the smallest
    possible value of n.   (OCR)

6 A particular brand of petrol was used in 80 randomly chosen cars of the same model and age. The petrol consumption, x miles per gallon, was obtained for each car. The results are summarised by $\sum x = 1896$ and $\sum x^2 = 45\,959$. Calculate an approximate symmetric 98% confidence interval for the mean petrol consumption of all cars of this model and age.

Give a reason why the interval is approximate. (OCR, adapted)

7 The random variable X has a normal distribution with unknown mean μ and known variance σ^2. A symmetric 98% confidence interval for μ, based on a random sample of 25 observations of X, is $[12.4, 12.6]$. Find

(a) the value of σ,

(b) the size of a random sample that will give a symmetric 95% confidence interval for μ of width as close as possible to 0.8.

8 A random sample of 200 fish was collected from a pond containing a large number of fish. Each was marked and returned to the pond. A day later, 400 fish were collected and 22 were found to be marked. These were also returned to the pond.

(a) Obtain a symmetric 90% confidence interval for the proportion of marked fish in the pond.

(b) Assuming that the number of marked fish is still 200, what can be said about the size of the population of fish in the pond?

9 The compiler of crossword puzzles classifies a puzzle as 'easy' if 60% or more people attempting the puzzle can complete it correctly within 20 minutes. It is classified as 'hard' if fewer than 30% of people can complete it correctly within 20 minutes. All other puzzles are classified as 'average'. A particular puzzle was given to 150 competitors in a contest and 74 completed it correctly within 20 minutes. The compiler wishes to be 90% confident of correctly classifying the puzzle.

(a) How should she classify the puzzle?

(b) Can she be 95% confident that her classification is correct?

10 A television viewer notices that many TV programmes start a few minutes later than the advertised time. For a random sample of 120 programmes, unbiased estimates of the mean and variance of the delay times are 1.75 minutes and 2.46 minutes^2 respectively, each correct to 3 significant figures. Use these estimates to calculate a symmetric 99% confidence interval for the population mean delay.

The viewer considers a normal distribution $N(1.75, 2.46)$ as a model for the delay, in minutes, at the start of a randomly chosen TV programme. Show that this model is not consistent with the viewer's observation that TV programmes almost never start before the advertised time.

Does the fact that the distribution of the delay times may not be normal affect the validity of the calculation of the confidence interval for the mean delay? Give a reason for your answer. (OCR)

6 Hypothesis testing: continuous variables

Part of the purpose of statistics is to help you to make informed decisions based on data. This chapter is the first of a series of chapters about decision making, which in statistics is called 'hypothesis testing'. When you have completed it you should

- understand the nature of a hypothesis test
- understand the difference between a one-tail and a two-tail test
- be able to formulate a null hypothesis and an alternative hypothesis
- understand the terms 'significance level', 'rejection region', 'acceptance region' and 'test statistic'
- be able to carry out a hypothesis test of a population mean for a sample drawn from a normal distribution of known variance, and also for a large sample.

6.1 An introductory example

Over a number of years a primary school has recorded the reading ages of children at the beginning and end of each academic year. The teachers have found that during Year 3 (the children's third year of school) the increase in reading age is normally distributed with mean 1.14 years and standard deviation 0.16 years. This year they are going to trial a new reading scheme: other schools have tried this scheme and found that it led to a greater increase in reading age. At the end of the year the teachers will use the mean increase in reading age, $\bar{x}$, of the 40 children in Year 3 to help them answer the question: 'Does the new reading scheme give better results than the old one in our school?'

The difficulty in answering this question lies in the fact that each child progresses at a different rate so that different values of $\bar{x}$ will be obtained for different groups of children. It is easy to check whether $\bar{x}$ is greater than 1.14 years. It is not easy to know whether a value of $\bar{x}$ greater than 1.14 years reflects the effectiveness of the new scheme or is just due to random variation between children.

This chapter describes a statistical method for arriving at a decision. The following sections break down the process into several stages.

6.2 Null and alternative hypotheses

There are two theories about how the new reading scheme performs in this particular school. The first is that using the new scheme makes no difference. This theory is called a **null hypothesis**. It is denoted by the symbol H_0. In this example, where the mean of the past increases in reading age is 1.14 years, the null hypothesis can be expressed by $H_0: \mu = 1.14$. Note that H_0 proposes a single value for the population mean, μ, which is based on past experience.

A 'hypothesis' is a theory which is assumed to be true unless evidence is obtained which indicates otherwise. 'Null' means 'nothing' and the term 'null hypothesis' means a 'theory of no change', that is 'no change' from what would be expected from past experience.

The other theory is that the new reading scheme is more effective than the old one, that is that the population mean will increase. This is called the **alternative hypothesis** and is given the symbol H_1. So the alternative hypothesis in this case is $H_1: \mu > 1.14$. The alternative hypothesis proposes the way in which μ will have changed if the new reading scheme is more effective than the old one.

The procedure which is used to decide between these two opposing theories is called a **hypothesis test** or sometimes a **significance test**. In this example the test will be **one-tail** because the alternative hypothesis proposes a change in the mean in only one direction, in this case an increase. It is also possible to have a one-tail test in which the alternative hypothesis proposes a decrease in the mean. Tests in which the alternative hypothesis suggests a difference in the mean in either direction are called **two-tail**.

Example 6.2.1
For the following situations give null and alternative hypotheses and say whether a hypothesis test would be one-tail or two-tail.
(a) In the past an athlete has run 100 metres in 10.3 seconds on average. He has been following a new training program which he hopes will decrease the time he takes to run 100 metres. He is going to time himself on his next six runs.
(b) The bags of sugar coming off a production line have masses which vary slightly but which should have a mean value of 1.01 kg . A sample is to be taken in order to test whether there has been any change in the mean.
(c) The mean volume of liquid in bottles of lemonade should be at least 2 litres. A sample of bottles is taken in order to test whether the mean volume has fallen below 2 litres.

> (a) The null hypothesis proposes a single value for μ, $H_0: \mu = 10.3$ based on the athlete's past performance. The alternative hypothesis proposes how μ might have changed, $H_1: \mu < 10.3$. This is a one-tail test.

> (b) The null hypothesis proposes a single value for μ, $H_0: \mu = 1.01$ based on the mass which the bags should have. The alternative hypothesis proposes how μ might have changed, $H_1: \mu \neq 1.01$. This is a two-tail test.

> (c) The only concern in this situation is whether the mean volume is less than 2 litres so a one-tail test is appropriate. The null hypothesis is $H_0: \mu = 2$ and the alternative hypothesis is $H_1: \mu < 2$.

For a hypothesis test on the population mean, μ, the null hypothesis, H_0, proposes a value, μ_0, for μ,

$$H_0: \mu = \mu_0.$$

The alternative hypothesis, H_1, suggests the way in which μ might differ from μ_0. H_1 can take three forms:

$H_1: \mu < \mu_0$, a one-tail test for a decrease;
$H_1: \mu > \mu_0$, a one-tail test for an increase;
$H_1: \mu \neq \mu_0$, a two-tail test for a difference.

Exercise 6A

In the following situations, state suitable null and alternative hypotheses involving a population with mean μ. You will need some of your answers in Exercise 6B.

1 Bars of Choco are claimed by the manufacturer to have a mean mass of 102.5 grams. A test is carried out to see whether the mean mass of Choco bars is less than 102.5 grams.

2 The mean factory assembly time for a particular electronic component is 84 s. It is required to test whether the introduction of a new procedure results in a different assembly time.

3 In a report it was stated that the average age of all hospital patients was 53 years. A newspaper believes that this figure is an underestimate.

4 The manufacturer of a certain battery claims that it has a mean life of 30 hours. A suspicious customer wishes to test the claim.

5 A large batch of capacitors is judged to be satisfactory by an electronics factory if the mean capacitance is at least 5 microfarads. A test is carried out on a batch to determine whether it is satisfactory.

6.3 Critical values

Once you have decided on your null and alternative hypotheses the next step is to devise a rule for choosing between them. Look again at the reading scheme example. The rule will be based on the sample mean, $\bar{x}$. The teachers are only interested in the new scheme if it improves the average increase in reading age and so only values of $\bar{x}$ greater than 1.14 might lead them to drop the old scheme in favour of the new one. Initially, you might think that *any* value of $\bar{x}$ which is greater than 1.14 years would show that the new scheme is more effective. A little more thought shows that is too simple a rule.

It is possible to obtain a sample mean $\bar{x}$ that is greater than 1.14 even if the new reading scheme is not effective at all. To see this, suppose that there is no difference between the new reading scheme and the old one. Then both μ and σ will be the same under the two schemes and X will be distributed as $N(1.14, 0.16^2)$.

You should recall from Section 4.6 that if

$X \sim N(\mu, \sigma^2)$, then $\bar{X} \sim N\left(\mu, \dfrac{\sigma^2}{n}\right)$; so for

samples consisting of 40 children, the mean,

$\bar{X}$, will be distributed as $N\left(1.14, \dfrac{0.16^2}{40}\right)$.

Fig. 6.1 shows the distribution of $\bar{X}$. You can see from this diagram that there is a probability of $\frac{1}{2}$ that the sample mean will be greater than 1.14 even though you assumed that there is no change in the population mean.

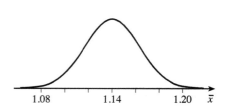

Fig. 6.1. Distribution of mean increase in reading age for the children in Section 6.1 if the new scheme is ineffective.

How big does the sample mean have to be before you can conclude that the population mean is likely to have increased from 1.14? Most people would agree that if a sample mean of 2.00 is obtained then it is unlikely that the population mean is still 1.14 but what about a sample mean of 1.19? One way of tackling this problem is to divide the possible outcomes into two regions: the **rejection** (or **critical**) **region** and the **acceptance region.**

The rejection region will contain values at the top end of the distribution in Fig. 6.1.

If the sample mean is in the rejection region, you reject H_0 in favour of H_1: you conclude that the population mean has increased. If the sample mean is in the acceptance region, you do not reject H_0: there is insufficient evidence to say that the new reading scheme is more effective.

The rejection region is chosen so that it is 'unlikely' for the sample mean to fall in the rejection region when H_0 is true. It is a matter of opinion what you mean by 'unlikely' but the usual convention among statisticians is that an event which has a probability of 0.05, that is 1 in 20, or less is 'unlikely'.

Fig. 6.2 shows the rejection region and the acceptance region for the children's reading scheme example. The value, c, which separates the rejection and acceptance regions is called a **critical value**. It can be calculated as follows.

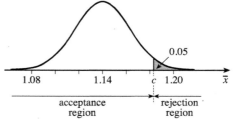

Fig. 6.2. Acceptance and rejection regions for the children's reading age example.

Since $\Phi(z) = 0.95$, the value of the standardised normal variable, z, corresponding to c, is $z = 1.645$ (using the tables on page 165).

The values of the standardised and original variables are related by $z = \dfrac{\bar{x} - \mu}{\sqrt{\dfrac{\sigma^2}{n}}}$, where $Z \sim N(0,1)$.

Substituting in this equation gives $1.645 = \dfrac{c - 1.14}{\sqrt{\dfrac{0.16^2}{40}}}$.

Rearranging gives $c = 1.645 \sqrt{\dfrac{0.16^2}{40}} + 1.14 = 1.18$, correct to 3 significant figures.

The rejection region is given by $\bar{X} \geqslant 1.18$ years.

At the end of the year the observed value for the sample mean was $\bar{x} = 1.19$ years. Since this value is in the rejection region, you can conclude that the observed result is unlikely to be explained by random variation: it is more likely to be due to an increase in the population mean. This suggests that the new reading scheme does give better results than the old one.

In this example a decision is made by considering the value of the sample mean. The sample mean, $\bar{X}$, is called the **test statistic** for this hypothesis test. The rejection region was defined so that the probability of the test statistic falling in it, *if H_0 is true*, is at most 0.05, or 5%.

This probability is called the **significance level** of the test. It gives the probability of rejecting H_0 when it is in fact true. In this example it gives the probability of concluding that the new reading scheme is better even when it is not. You might feel that this is too high a risk of being wrong and choose instead to use a significance level of, say, 0.01, or 1%.

Example 6.3.1
Find the rejection region for a test at the 1% significance level for the children's reading scheme example.

Now $\Phi(z) = 0.99$ giving $z = 2.326$. Substituting into $z = \dfrac{\bar{x} - \mu}{\sqrt{\dfrac{\sigma^2}{n}}}$ gives

$$2.326 = \frac{c - 1.14}{\sqrt{\dfrac{0.16^2}{40}}}.$$

Rearranging gives $c = 1.20$, correct to 3 significant figures.

The rejection region is $\overline{X} \geqslant 1.20$.

The observed value of 1.19 years is no longer in the rejection region and so H_0 is not rejected at the 1% significance level.

You may feel that it is unsatisfactory that the result of a hypothesis test should depend on the significance level chosen. This point is discussed in more detail in Chapter 8.

In a two-tail test the rejection region has two parts, because both high and low values of $\overline{X}$ are unlikely if the null hypothesis is true. Example 6.3.2 illustrates this situation.

Example 6.3.2
In the past a machine has produced rope which has a breaking load which is normally distributed with mean 1000 N and standard deviation 21 N. A new process has been introduced. In order to test whether the mean breaking load has changed a sample of 50 pieces of rope is taken, the breaking strain of each piece measured and the mean calculated.
(a) Define suitable null and alternative hypotheses for testing whether the breaking load has changed.
(b) Taking the sample mean as the test statistic, find the rejection region for $\overline{X}$ for a hypothesis test at the 5% significance level.
(c) The sample mean for the 50 pieces of rope was 1003 N. What can you deduce?

(a) The null hypothesis states the value which the mean breaking load should take, $H_0 : \mu = 1000$.

The alternative hypothesis states how μ might have changed, $H_1 : \mu \neq 1000$.

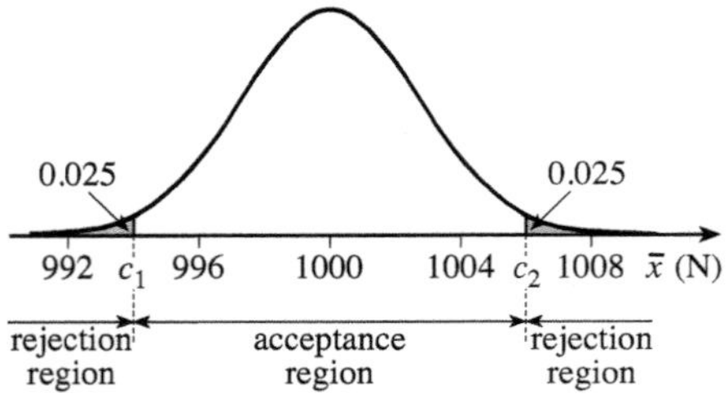

(b) If H_0 is true, then the sample mean,

$\overline{X} \sim N\left(1000, \dfrac{21^2}{50}\right)$. Fig. 6.3 shows the

distribution of $\overline{X}$ with the rejection and acceptance regions. There are two critical values labelled c_1 and c_2.

Fig. 6.3. Acceptance and rejection regions for Example 6.3.2.

To find the upper critical value, use $\Phi(z) = 0.975$, since the 0.05 probability is split equally between the two 'tails' of the distribution. The required value of z is 1.960.

Substituting into $z = \dfrac{\bar{x} - \mu}{\sqrt{\dfrac{\sigma^2}{n}}}$ gives $1.960 = \dfrac{c_2 - 1000}{\sqrt{\dfrac{21^2}{50}}}$.

Rearranging gives $c_2 = 1006$, to the nearest integer. By symmetry, $c_1 = 994$.

So the rejection region is $\bar{X} \leqslant 994$ and $\bar{X} \geqslant 1006$.

(c) The observed sample mean of $\bar{x} = 1003$ is not in the rejection region. There is not enough evidence to say that the mean has changed and it can be concluded that the new process is satisfactory.

Here is a summary of the terms introduced in this section, followed by a list of the steps involved in carrying out a hypothesis test.

> The **test statistic** is calculated from the sample. Its value is used to decide whether the null hypothesis, H_0, should be rejected.
>
> The **rejection** (or **critical**) **region** gives the values of the test statistic for which the null hypothesis, H_0, is rejected.
>
> The **acceptance region** gives the values of the test statistic for which the null hypothesis, H_0, is not rejected.
>
> The boundary value(s) of the rejection region is (are) called the **critical value(s)**.
>
> The **significance level** of a test gives the probability of the test statistic falling in the rejection region when H_0 is true.

If H_0 is rejected, then H_1 is automatically accepted.

> To carry out a hypothesis test:
>
> **Step 1** Define the null and alternative hypotheses.
>
> **Step 2** Decide on a significance level.
>
> **Step 3** Determine the critical value(s).
>
> **Step 4** Calculate the test statistic.
>
> **Step 5** Decide on the outcome of the test depending on whether the value of the test statistic is in the rejection or the acceptance region.
>
> **Step 6** State the conclusion in words.

Exercise 6B

In the following questions, the rejection (critical) regions should be found in terms of the sample mean, $\overline{X}$.

1 The random variable X has a normal distribution, $N(\mu,4)$. A test of the null hypothesis $\mu = 10$ against the alternative hypothesis $\mu > 10$ is carried out, at the 5% significance level, using a random sample of 9 observations of X. The rejection region is found to be $\overline{X} \geqslant 11.10$. State the conclusion of the test in the following cases.

    (a) $\overline{X} = 12.3$ (b) $\overline{X} = 8.6$

2 For the situation in Exercise 6A Question 1, a random sample of 12 bars had a mean mass of 101.4 g. Test, at the 5% significance level, whether the mean mass of all Choco bars is less than 102.5 g, assuming that the mass of a Choco bar is normally distributed with standard deviation 1.7 g.

3 For the situation in Exercise 6A Question 2, a random sample of 40 components had mean assembly time 81.2 s. Assuming that the assembly time of a component has a normal distribution with standard deviation 6.1 s, carry out a test at the 5% significance level of whether the mean for all components differs from 84.0 s.

4 Referring to Exercise 6A Question 5, a random sample of 6 capacitors was selected from the batch. Their capacitances were measured in microfarads with the following results.

    5.12 4.81 4.79 4.85 5.04 4.61

Assuming that the capacitances have a normal distribution with standard deviation 0.35 microfarads, test, at the 2% significance level, whether the batch is satisfactory.

5 The blood pressure of a group of hospital patients with a certain type of heart disease has mean 85.6. A random sample of 25 of these patients volunteered to be treated with a new drug and a week later their mean blood pressure was found to be 70.4. Assuming a normal distribution with standard deviation 15.5 for blood pressures, and using a 1% significance level, test whether the mean blood pressure for all patients treated with the new drug is less than 85.6.

6 Two-litre bottles of a brand of spring water are advertised as containing 6.8 mg of magnesium. In a random sample of 10 of these bottles the mean amount of magnesium was found to be 6.92 mg. Assuming that the amounts of magnesium are normally distributed with standard deviation 0.18 mg, test whether the mean amount of magnesium in all similar bottles differs significantly from 6.8 mg. Use a 5% significance level.

7 The lives of a certain make of battery have a normal distribution with mean 30 h and variance 2.54 h^2. When a large consignment of these batteries is delivered to a store the quality control manager tests the lives of 8 randomly chosen batteries. The mean life was 28.8 h. Test whether there is cause for complaint. Use a 3% significance level.

8 The birth weights of babies born in a certain large hospital maternity unit during the year 2000 had a normal distribution with mean 3.21 kg and standard deviation 0.73 kg. During the first week of August, there were 24 babies born with a mean weight of 3.17 kg. Treating these 24 babies as a random sample, would a test at the 5% significance level indicate a population mean which is different from 3.21 kg?

6.4 Standardising the test statistic

In the previous exercise the rejection region for each question was different and you had to find it before you could obtain the result of the hypothesis test. You may have spotted that the calculation could be shortened by standardising the value of $\overline{X}$ using

$$Z = \frac{\overline{X} - \mu}{\sqrt{\dfrac{\sigma^2}{n}}} = \frac{\overline{X} - \mu}{\dfrac{\sigma}{\sqrt{n}}}$$

and taking Z as the test statistic. For a given type of test, one-tail or two-tail, at a given significance level the rejection region for Z will always be the same. For example, Fig. 6.4 illustrates the rejection region of Z for a two-tail test at the 5% significance level. The upper critical value is obtained from $\Phi(z_1) = 0.975$, giving $z_1 = 1.960$ and, by symmetry, the lower critical value, $z_2 = -1.960$. Thus the rejection region is $Z \geqslant 1.960$ and $Z \leqslant -1.960$, which you can write more compactly as $|Z| \geqslant 1.960$. The following examples illustrate this approach.

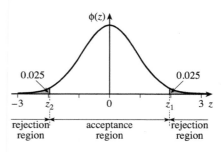

Fig. 6.4. Acceptance and rejection regions for Z for a two-tail test at the 5% significance level.

Example 6.4.1

A test of mental ability has been constructed so that, for adults in Great Britain, the test score is normally distributed with mean 100 and standard deviation 15. A doctor wishes to test whether sufferers from a particular disease differ in mean from the general population in their performance on this test. She chooses a random sample of 10 sufferers. Their scores on the test are

$$119 \quad 131 \quad 95 \quad 107 \quad 125 \quad 90 \quad 123 \quad 89 \quad 103 \quad 103.$$

Carry out a test at the 5% significance level to test whether sufferers from the disease differ from the general population in the way in which they perform at this test.

(OCR, adapted)

The null and alternative hypotheses are $H_0: \mu = 100$ and $H_1: \mu \neq 100$ respectively.

This is a two-tail test at the 5% significance level. As explained above, the rejection region for the test statistic, Z, is $|Z| \geqslant 1.960$.

Under H_0, $\overline{X} \sim N\left(100, \dfrac{15^2}{10}\right)$.

'Under H_0' is another way of saying 'If H_0 is true'.

For this sample,

$$\bar{x} = \tfrac{1}{10}(119 + 131 + 95 + 107 + 125 + 90 + 123 + 89 + 103 + 103) = 108.5.$$

When $\bar{x} = 108.5$, $z = \dfrac{\bar{x} - \mu}{\sqrt{\dfrac{\sigma^2}{n}}} = \dfrac{(108.5 - 100)}{\sqrt{\dfrac{15^2}{10}}} = 1.792$, correct to 3 decimal places.

The observed value of Z is not in the rejection region so H_0 is not rejected. There is insufficient evidence, at the 5% significance level, to suggest that sufferers from this disease differ from the general population in their performance on the test.

Example 6.4.2

A manufacturer claims that its light bulbs have a lifetime which is normally distributed with mean 1500 hours and standard deviation 30 hours. A shopkeeper suspects that the bulbs do not last as long as is claimed because he has had a number of complaints from customers. He tests a random sample of six bulbs and finds that their lifetimes are 1472, 1486, 1401, 1350, 1511, 1591 hours. Is there evidence, at the 1% significance level, that the bulbs last a shorter time than the manufacturer claims?

The null and alternative hypotheses are $H_0 : \mu = 1500$ and $H_1 : \mu < 1500$ respectively.

This is a one-tail test for a decrease at the 1% level. Fig. 6.5 shows the rejection region for Z. The critical value is given by $\Phi(z) = 0.01$. Recall that $\Phi(-z) = 1 - \Phi(z) = 1 - 0.01 = 0.99$. From tables, $-z = 2.326$ so $z = -2.326$. Thus the rejection region is $Z \leqslant -2.326$.

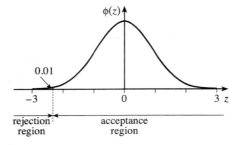

Fig. 6.5. Acceptance and rejection regions for Example 6.4.2.

Under H_0, $\overline{X} \sim N\left(1500, \dfrac{30^2}{6}\right)$.

For this sample,

$$\bar{x} = \tfrac{1}{6}(1472 + 1486 + 1401 + 1350 + 1511 + 1591) = 1468.5.$$

When $\bar{x} = 1468.5$, $z = \dfrac{\bar{x} - \mu}{\sqrt{\dfrac{\sigma^2}{n}}} = \dfrac{1468.5 - 1500}{\sqrt{\dfrac{30^2}{6}}} = -2.572$, correct to 3 decimal places.

The observed value of Z is in the rejection region. There is evidence, at the 1% significance level, that the manufacturer's bulbs do not last as long as claimed.

You can generalise this method as follows.

The test statistic Z can be used to test a hypothesis about a population mean, $H_0: \mu = \mu_0$, for samples drawn from a normal distribution of known variance σ^2. For a sample of size n, the value of Z is given by

$$z = \frac{\bar{x} - \mu}{\sqrt{\dfrac{\sigma^2}{n}}} = \frac{\bar{x} - \mu}{\dfrac{\sigma}{\sqrt{n}}}. \qquad (6.1)$$

The rejection region for Z depends on H_1 and the significance level used. The critical values for some commonly used rejection regions are given below.

Significance level	Two-tail $H_1: \mu \neq \mu_0$	One-tail $H_1: \mu > \mu_0$	One-tail $H_1: \mu < \mu_0$
10%	±1.645	1.282	−1.282
5%	±1.960	1.645	−1.645
2%	±2.326	2.054	−2.054
1%	±2.576	2.326	−2.326

For some significance levels, critical values for a one-tail test for an increase in the population mean can be found from the table at the bottom of page 165. The probabilities given refer to the acceptance region. For the remaining significance levels, the critical values can be found using the full normal distribution table 'in reverse'.

Exercise 6C

1 Cans of lemonade are filled by a machine which is set to dispense an amount which is normally distributed with mean 330 ml and standard deviation 2.4 ml. A quality control manager suspects that the machine is over-dispensing and tests a random sample of 8 cans. The volumes of the contents, in ml, are as follows.

    329   327   331   326   334   343   328   339

Test, at the $2\frac{1}{2}$% significance level, whether the manager's suspicion is justified.

2 The masses of loaves from a certain bakery have a normal distribution with mean μ grams and standard deviation σ grams. When the baking procedure is under control, $\mu = 508$ and $\sigma = 18$. A random sample of 25 loaves from a day's output had a total mass of 12 554 grams. Does this provide evidence at the 10% significance level that the process is not under control?

3 A machine produces elastic bands with breaking tension T newtons, where $T \sim N(45.1, 19.0)$. On a certain day, a random sample of 50 bands was tested and found to have a mean breaking tension of 43.4 newtons. Test, at the 4% significance level, whether this indicates a change in the mean breaking tension.

4 The cholesterol level of healthy males under the age of 21 is normally distributed with mean 160 and standard deviation 10. A random sample of 200 university students, all male and under age 21, had a mean cholesterol level of 161.8. Test, at the 1% significance level, whether all male university students under age 21 have a mean cholesterol level greater than 160.

5 The mean and standard deviation of the number of copies of *The Daily Courier* sold by a shop were 276.4 and 12.2 respectively. During 24 days following an advertising campaign, the total number of copies of *The Daily Courier* sold by the shop was 6713. Stating your assumptions, test at the 5% significance level whether the data indicate that the campaign was successful.

6 The average time that I have to wait for the 0815 bus is 4.3 minutes. A new operator takes over the service, with the same timetable, and my average waiting time for 10 randomly chosen days under the new operator is 3.4 minutes. Assuming that the waiting time has a normal distribution with standard deviation 1.8 minutes, test whether the average waiting time under the new operator has decreased. Use a 10% significance level.

7 The marks of all candidates in an A-Level Statistics examination were normally distributed with mean 42.3 and standard deviation 11.2. Fifteen students from a particular school sat this paper as a practice examination. Their mean mark was 49.8. Test, at the 1% significance level, whether this indicates that the students at this school did better than candidates in general.

6.5 Large samples

In the examples considered so far you have assumed that the samples were drawn from a normal distribution and that the variance of this distribution is known. In practice, there are many cases where you cannot be sure that the population distribution is normal and you may or may not have accurate information about its variance. However, the method of hypothesis testing which has been described in this chapter can still be applied provided that the sample is large. Consider first the distribution from which the sample is taken. Although this may not be normal, the central limit theorem (see Section 4.4) tells you that the sample mean, $\overline{X}$, is normally distributed provided that the sample is large. Next, consider the population variance. An unbiased estimate of this can be calculated from the sample using the equation on page 81.

$$s^2 = \frac{1}{n-1}\left(\sum x^2 - \frac{(\sum x)^2}{n}\right).$$

If the sample is large, this estimate is sufficiently accurate to replace σ^2 in Equation 6.1.

> The test statistic Z can be used to test a hypothesis about a population mean, $H_0: \mu = \mu_0$, for large samples drawn from any population.
>
> For a sample of size n, the value of Z is given by Equation 6.1.
> If the population variance, σ^2, is unknown, it can be replaced by its estimate, s^2.

The term 'large' is not very precise. A rule of thumb would be that 'large' means a sample size of 30 or more.

Example 6.5.1

A new surgical technique has been developed in an attempt to reduce the time that patients have to spend in hospital after a particular operation. In the past, the mean time spent in hospital was 5.3 days. For the first 50 patients on whom the new technique was tried, the mean time spent in hospital was 5.0 days with an estimated population variance of 0.4^2 days^2. Is there evidence at the 2% significance level that the new technique has reduced the time spent in hospital?

The null and alternative hypotheses are $H_0: \mu = 5.3$ and $H_1: \mu < 5.3$ respectively.

The population distribution is not given, but, since the sample is large, the test statistic Z can still be used. This is a one-tail test for a decrease at the 2% significance level. From the box on page 111 the rejection region is $Z \leqslant -2.054$.

$$z = \frac{\bar{x} - \mu}{\sqrt{\dfrac{s^2}{n}}} = \frac{5.0 - 5.3}{\sqrt{\dfrac{0.4^2}{50}}} = -5.303, \text{ correct to 3 decimal places.}$$

The calculated value of z is in the rejection region so H_0 is rejected: there is evidence at the 2% significance level that the time in hospital has been reduced by the new technique.

Example 6.5.2

An inspector of items from a production line takes, on average, 21.75 seconds to check each item. After the installation of a new lighting system the times, t seconds, to check each of 50 randomly chosen items from the production line are summarised by $\sum t = 1107$ and $\sum t^2 = 24\,592.35$.

(a) Calculate an unbiased estimate of the population variance of the time to check an item under the new lighting system.
(b) Test at the 5% significance level whether there is evidence that the population mean time has changed from 21.75 seconds.   (OCR, adapted)

(a) An unbiased estimate of population variance is given by

$$s^2 = \frac{1}{n-1}\left(\sum t^2 - \frac{\left(\sum t\right)^2}{n}\right) = \frac{1}{49}\left(24\,592.35 - \frac{1107^2}{50}\right) = 1.701\ldots\,.$$

(b) The null and alternative hypotheses are $H_0: \mu = 21.75$ and $H_1: \mu \neq 21.75$.

The population distribution and variance are not given, but, since the sample is large, the test statistic Z can still be used. For a two-tail test at the 5% significance level the rejection region for the test statistic Z is $|Z| \geqslant 1.960$.

For the given sample, $\bar{x} = \frac{1107}{50} = 22.14$.

$$z = \frac{\bar{x} - \mu}{\sqrt{\frac{s^2}{n}}} = \frac{22.14 - 21.75}{\sqrt{\frac{1.701\ldots}{50}}} = 2.114 \text{, correct to 3 decimal places.}$$

This value is in the rejection region. There is evidence at the 5% significance level that the population mean has changed.

Exercise 6D

1 The continuous random variable X has mean μ. A test of the hypothesis $\mu = 23$ is to be carried out at the 2% significance level. A random sample of 50 observations of X gave a sample mean $\bar{x} = 21.8$ with an estimated population variance $s^2 = 12.94$. Do the test.

2 A machine set to produce metal discs of diameter 11.90 cm has an annual service. After the service a random sample of 36 discs is measured and found to have a mean diameter of 11.930 cm and an estimated population variance of 0.072 cm^2. Test, at the 1% significance level, whether the machine is now producing discs of mean diameter greater than 11.90 cm.

3 In Lanzarote the mean daily number of hours of sunshine during April is reported to be $5\frac{1}{4}$ hours. During a particular year, during April, the daily amounts of sunshine, x hours, were recorded and the results are summarised by $\sum x = 162.3$ and $\sum x^2 = 950.6$. Test, at the 5% significance level, whether these results indicate that the reported mean of $5\frac{1}{4}$ hours is too low.

4 Boxes of the breakfast cereal Crispo indicate that they contain 375 grams. After receiving several complaints that the boxes contain less than the stated amount, a supermarket manager weighs the contents of a random sample of 40 boxes from a large consignment. The masses, x grams, of the contents are summarised by $\sum (x - 375) = -46$ and $\sum (x - 375)^2 = 616$. Test, at the 5% significance level, whether the mean mass of the contents of all boxes in the consignment is less than 375 grams.

5 The Galia melons produced by a fruit grower under usual conditions have a mean mass of 0.584 kg. The fruit grower decides to produce a crop organically and a random sample of 75 melons, ready for market, had masses, x kg, summarised by $\sum x = 45.39$ and $\sum x^2 = 29.03$. Test, at the 10% significance level, whether melons grown organically are heavier, on average, than those grown under the usual conditions.

6 The diameters of a random sample of 60 cans of a certain brand of tomato were measured. The results, x cm, are summarised by $\sum(x-6) = 48.9$ and $\sum(x-6)^2 = 40.07$. The mean diameter of all the cans produced is denoted by μ cm. Test, at the $2\frac{1}{2}\%$ significance level, the following hypotheses:

(a) $\mu \neq 6.8$,   (b) $\mu > 6.8$,   (c) $\mu < 6.8$.

6.6 An alternative method of carrying out a hypothesis test

Another way of carrying out a hypothesis test is to calculate the probability that the test statistic takes the observed value (or a more extreme value) and to compare this probability with the significance level. If the probability is less than the significance level then the null hypothesis is rejected. The result is said to be 'significant' at the given significance level. Fig. 6.6 shows that this method will always give the same result as the previous method.

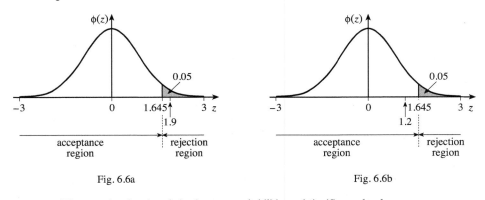

Fig. 6.6a   Fig. 6.6b

Diagrams showing the relation between probabilities and significance levels.

Fig. 6.6a shows the rejection region for Z for a one-tail test for an increase at the 5% significance level. If Z takes a value in the rejection region, for example 1.9, then you can see from this figure that $P(Z \geq 1.9)$ is less than 0.05, which would also lead to the rejection of H_0. If Z takes a value in the acceptance region, for example 1.2, as shown in Fig. 6.6b, then $P(Z \geq 1.2)$ is greater than 0.05 and H_0 would not be rejected. To illustrate this idea, look again at Example 6.5.1. In this example, the test statistic took the value -5.30 (correct to 3 significant figures). Low values of Z were of interest so a 'more extreme value' here means a value less than -5.30. The probability that Z took this value or a more extreme value is

$$P(Z \leq -5.30) = 1 - \Phi(5.30) \approx 0.$$

Since this probability is less than 2%, the result is significant at the 2% level and so the null hypothesis is rejected. As you would expect this is the same as the conclusion which was reached before. However, this way of quoting the result conveys more information: it shows that the result was significant not only at the 2% level but at a much lower significance level. Giving a probability (sometimes called a p-value) rather than using critical values requires a little more work. However, the use of microcomputers means that it is now easy to give a probability and most statistical programs give the result of a hypothesis test in this form.

The following example illustrates this approach in a two-tail test.

Example 6.6.1

A machine is designed to produce rods 2 cm long with a standard deviation of 0.02 cm. The lengths may be taken as normally distributed. The machine is moved to a new position in the factory, and in order to check whether the setting for the mean length has altered, the lengths of the first ten rods are measured. The standard deviation may be considered to be unchanged. If these lengths, in cm, are as given below, test at the 5% significance level whether the setting has altered or not.

$$2.04 \quad 1.97 \quad 1.99 \quad 2.03 \quad 2.04 \quad 2.10 \quad 2.01 \quad 1.98 \quad 1.97 \quad 2.02 \qquad \text{(OCR, adapted)}$$

This is a two-tail test with the null hypothesis assuming that the mean is unaltered.

The null and alternative hypotheses are $H_0: \mu = 2$ and $H_1: \mu \neq 2$ respectively.

$$\text{Sample mean} = \tfrac{1}{10}(2.04 + 1.97 + 1.99 + 2.03 + 2.04 + 2.10 + 2.01 + 1.98 + 1.97 + 2.02)$$
$$= 2.015.$$

Since the population is normally distributed, $\overline{X}$ is also normally distributed.

Under H_0, $\overline{X} \sim N\left(2, \dfrac{0.02^2}{10}\right)$.

$$\text{Thus } P(\overline{X} \geqslant 2.015) = P\left(Z \geqslant \frac{2.015 - 2}{\dfrac{0.02}{\sqrt{10}}}\right) = P(Z \geqslant 2.372)$$

$$= 1 - \Phi(2.372) = 1 - 0.9912 = 0.0088 = 0.88\%.$$

Since this is a two-tail test, this probability should be compared with half of the value specified in the significance level, that is $2\tfrac{1}{2}\%$, as shown in Fig. 6.7. Since 0.88% is less than $2\tfrac{1}{2}\%$, the result is significant at the 5% level. It can be assumed that the mean length of the rods produced by the machine has been affected by the move.

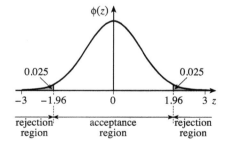

Fig. 6.7. Acceptance and rejection regions for a two-tail test at the 5% significance level.

Alternatively, you could calculate the p-value for the test. This is the probability that the test statistic deviates from the population mean by the observed amount (or more) *in either direction*. Thus, in this case, the p-value is $2 \times 0.88\% = 1.76\%$. The p-value is then compared with the significance level. Since $1.76\% < 5\%$, the result is significant at the 5% level, as before.

Either method of carrying out a hypothesis test, using critical values or using probabilities, is satisfactory and usually you should use the method which you find easier. However, in the next chapter you will meet situations where the probability approach is simpler. For this reason, it is suggested that you carry out Exercise 6E using the probability method.

6.7 Practical activities

1 Just a minute! *(You may already have data available for this from S1 Section 3.12.)*

Do people tend to over- or underestimate time intervals? Ask as large a sample of people as possible to estimate a time interval of one minute. You will need to decide on a standard procedure for doing this. Record the value of the estimates to the nearest second. Use your sample to test the null hypothesis $H_0 : \mu = 60$. You could carry out this experiment for more than one distinct group, for example children in a certain age range and adults.

2 Investigating an optical illusion Each person in the sample is presented with a diagram similar to that shown in Fig. 6.8 and asked to mark the centre of the horizontal line by eye. Ask as many people as possible.

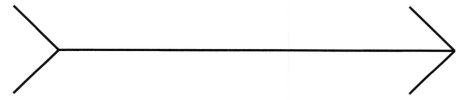

Fig. 6.8

Another sample is asked to mark the centre of a line of the same length as before but without the arrows on the ends.

Analyse the results as follows. Measure the deviation of each mark from the centre of the line, taking deviations to the right as positive and those to the left as negative. For each sample test whether the mean deviation differs significantly from zero. Comment on your results.

Exercise 6E

Use the method of Section 6.6 to answer the questions in this exercise.

1 The random variable X has a normal distribution with mean μ and variance 25. A random sample of 20 observations of X is taken and the sample mean is denoted by $\overline{X}$. This is used to test the null hypothesis $\mu = 30$ against the alternative hypothesis $\mu < 30$.

(a) Calculate $P(\overline{X} \leqslant 28.4)$.

(b) If the sample mean is, in fact, 28.4, state whether the null hypothesis is rejected at the

(i) 5% significance level,   (ii) 10% significance level.

2 A charity has a large number of collection boxes in a variety of locations. They are emptied on a regular basis and during 2000 the mean amount collected per box (on emptying) was $8.54. Early in 2001, a random sample of 50 boxes was emptied and the contents, $\$x$, are summarised by $\sum (x - 8) = 38.4$ and $\sum (x - 8)^2 = 240.41$. Find the p-value of a test of whether the data provides evidence that the mean is now different from $8.54.

What conclusion can be made at the 5% significance level?

3   The alkalinity of soil is measured by its pH value. It has been found from many previous measurements that the pH values in a particular area have mean 8.42 and standard deviation 0.74. After an unusually hot summer the pH values were measured at 36 randomly chosen locations in the area and the sample mean value was found to be 8.63. Calculate the p-value of a test of the hypothesis that the mean pH value in the area is now greater than 8.42.

What can be concluded

(a) at the 5% significance level,   (b) at the 1% significance level?

4   Longbrite candles are supposed to burn for at least 7 hours. To check this, a random sample was tested; 30 Longbrite candles were lit, and the times, t hours, before they went out were recorded. The results are summarised by $\sum(t-7) = -2.40$ and $\sum(t-7)^2 = 2.95$. Find the p-value of a test of whether the Longbrite candles burn, on average, for at least 7 hours.

What is the conclusion of the test at the 10% significance level?

5   The percentage salinity of the water in a stretch of sea was measured at 45 randomly chosen places during 2002. The sample mean percentage was 31.18 and an unbiased estimate of population variance was 0.579^2. Find the p-value of a test of whether the mean salinity of the sea in that area is greater than 31%. State the conclusion of the test at the 5% significance level.

6   A teacher writes an examination paper which she thinks the average student should take 80 minutes to complete. She gave the paper to 35 randomly chosen students. The sample mean was 81.4 minutes and an unbiased estimate of population variance was 2.9^2 min^2. Find the smallest significance level at which it would be accepted that the mean time for all students taking the paper would differ from 80 minutes.

Miscellaneous exercise 6

1   Metal struts used in a building are specified to have a mean length of 2.855 m. The lengths have a normal distribution with standard deviation 0.0352 m. A batch of 15 struts is sent to a building site and the lengths are measured. The sample mean length is 2.841 m.

A test is to be carried out, at the 5% significance level, to decide whether the batch is from the specified population.

(a) Stating your hypotheses, find the rejection region in terms of Z.

(b) State the conclusion of the test.

2   A random variable, X, has a normal distribution with unknown mean but known variance of 12.4. The mean of a random sample of 10 observations of X is denoted by $\overline{X}$. The acceptance region of a test of the null hypothesis $\mu = 25$ is $\overline{X} > 22.41$.

(a) State the alternative hypothesis.   (b) Find the significance level of the test.

(c) If the hypothesised value of μ were greater than 25 would the significance level corresponding to the same acceptance region be larger or smaller than that found in part (b)? Give a reason for your answer.

3 The contents of a brand of Greek yoghurt can be assumed to have masses which are normally distributed with standard deviation 2.58 grams. A new machine for filling the cartons is purchased and a random sample of 20 cartons filled by this machine is used to test the null hypothesis that the mean mass of the contents is μ_0 grams. Using the standard deviation of 2.58 grams, the rejection region of the test is $\overline{X} \leqslant 208.81$ or $\overline{X} \geqslant 211.19$.

(a) Find the value of μ_0.   (b) Find the significance level of the test.

4 Nisha, who has diabetes, has to monitor her blood glucose levels, which vary throughout the day. The results from a sample of 75 readings, x, taken at random times over a week, are summarised by $\sum x = 511.5$ and $\sum x^2 = 4027.89$.

(a) Assuming a normal distribution, test at the 5% significance level whether Nisha's mean blood glucose level, μ, is greater than 6.0.

(b) Find the set of values of μ_0 for which it would be accepted that $\mu > \mu_0$ at the 10% significance level.

(c) State, giving a reason, whether the conclusion of the test in part (a) would be valid

(i) if it could no longer be assumed that blood glucose level has a normal distribution,

(ii) if the 75 readings were all taken at week-ends.

5 A total crop weight, x kg, of each of 64 bean plants is measured by a horticulturalist and the results are summarised by $\sum x = 303.4$ and $\sum x^2 = 1615.96$. Find unbiased estimates of the population mean and variance.

The horticulturalist wishes to test the hypothesis that the mean crop weight per plant is 5 kg against the alternative hypothesis that the mean crop weight is less that 5 kg. Carry out the test at the 10% significance level.

Find the smallest significance level at which the test would result in rejection of the null hypothesis.   (OCR)

6 An athletics coach has the use of a gymnasium which he sets out for circuit training. The time taken for a new athlete to complete the circuit on each of a large number of occasions is noted. The mean value of these times is 100 seconds and the standard deviation is 3 seconds. The distribution of the times may be assumed to be normal. Three months later, after the athlete has been training daily, his times, in seconds, to complete the circuit on 10 occasions are as follows.

    96.8   101.2   98.2   99.6   98.0   95.6   98.0   100.0   95.2   97.4

Verify that these values show, at the 5% significance level, that the athlete's performance in circuit training has improved.

Calculations based on a large number of further trials show that at this stage the mean value of the time taken by the athlete is 98 seconds and the standard deviation is 3 seconds. What would be the maximum total time for 10 circuits at some later stage which would provide evidence, at the 5% significance level, of further improvement?   (OCR)

7 An ambulance station serves an area which includes more than 10 000 houses. It has been decided that if the mean distance of the houses from the ambulance station is greater than 10 miles then a new ambulance station will be necessary. The distance, x miles, from the station of each of a random sample of 200 houses was measured, the results being summarised by $\sum x = 2092.0$ and $\sum x^2 = 24\,994.5$.

(a) Calculate, to 4 significant figures, unbiased estimates of

   (i) the population mean distance, μ miles, of houses from the station,

   (ii) the population variance of the distance of the houses from the station.

   State what you understand by 'unbiased estimate'.

(b) A test of the null hypothesis $\mu = 10$ against the alternative hypothesis $\mu > 10$ is carried out at the $\alpha\%$ significance level, using a random sample of size 200. The rejection region for this test is $\overline{X} \geqslant 10.65$, where $\overline{X}$ denotes the sample mean.

   (i) Calculate the value of α.

   (ii) State the conclusion of the test using the sample data.

(c) Suppose that it could not be assumed that the distances are normally distributed. State whether the answers to part (a) and part (b) would still hold.   (OCR, adapted)

7 Hypothesis testing: discrete variables

This chapter takes further the idea of hypothesis testing introduced in the previous chapter. When you have completed it you should

- be able to formulate hypotheses and carry out a test of a population proportion by direct evaluation of binomial probabilities or by a normal approximation, as appropriate
- be able to formulate hypotheses and carry out a test of a population mean using a single observation drawn from a Poisson distribution, using either direct evaluation of probabilities or by a normal approximation, as appropriate.

7.1 Testing a population proportion

You may have seen advertisements for dairy spreads which claim that the spread cannot be distinguished from butter. How could you set about testing this claim? One way would be to take pairs of biscuits and put butter on one biscuit in each pair and the dairy spread on the other. The pairs of biscuits would be given to a number of tasters who would be asked to identify the biscuit with butter on it. Half the tasters would be given the buttered biscuit first, and the other half the buttered biscuit second.

Suppose you decided to use 10 tasters. How would you set about drawing a conclusion from your results? The method of hypothesis testing described in the last chapter can be adapted to this situation. First it is necessary to formulate a null hypothesis and an alternative hypothesis. It is usual to start from a position of doubt: you assume that the tasters cannot identify the butter and that they are guessing. In this situation the probability that a taster chosen at random will get the correct result is $\frac{1}{2}$. This can be expressed by the null hypothesis $H_0: p = \frac{1}{2}$. If some of the tasters can actually identify the butter then $p > \frac{1}{2}$. This can be expressed as an alternative hypothesis, $H_1: p > \frac{1}{2}$.

Can you see why it is difficult to take any other null hypothesis?

If H_0 is true, the number, X, of tasters who identify the buttered biscuit correctly is a random variable with distribution $B(10, \frac{1}{2})$. Fig. 7.1 shows this distribution. High values of X would suggest that H_0 should be rejected in favour of H_1. The most straightforward method of carrying out a hypothesis test for a discrete variable is to use the approach of Section 6.6 and calculate the probability that X takes the observed or a more extreme value assuming that H_0 is true and compare this probability with the specified significance level. Suppose that 9 out of the 10 people had identified the butter and you chose a significance level of 5%.

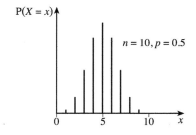

Fig. 7.1. Distribution $B(10, \frac{1}{2})$.

$$P(X \geqslant 9) = P(X = 9) + P(X = 10)$$
$$= \binom{10}{9}\left(\frac{1}{2}\right)^9\left(\frac{1}{2}\right) + \binom{10}{10}\left(\frac{1}{2}\right)^{10}$$
$$= 0.009\,76\ldots + 0.000\,976\ldots$$
$$= 0.010\,74\ldots$$
$$= 1.07\%, \text{correct to 3 significant figures.}$$

This probability is less than 5% so the result is significant at the 5% level. H_0 is rejected and there is evidence, at the 5% significance level, that the proportion of people who can distinguish the butter is greater than $\frac{1}{2}$.

As in the previous chapter the possible values of X can be divided into an acceptance region and a rejection region. However the situation here is complicated by the fact that X is a discrete variable. Table 7.2 shows the probability distribution of X.

x	$P(X = x)$
0	0.0010
1	0.0098
2	0.0439
3	0.1172
4	0.2051
5	0.2461
6	0.2051
7	0.1172
8	0.0439
9	0.0098
10	0.0010

Table 7.2. Probability distribution for $B\left(10, \frac{1}{2}\right)$.

For a significance level of, say, 5%, you will find that there is no rejection region which exactly corresponds to this probability. For example, for a rejection region of $X \geqslant 8$, the probability of a result in the rejection region is $0.0439 + 0.0098 + 0.0010 = 0.0547$ and so the actual significance level of the test is 5.47%; for $X \geqslant 9$, the actual significance level is 0.99%. This point will be considered in more detail in Section 8.3.

The following examples show further tests of this type.

Example 7.1.1
A national opinion poll claims that 40% of the electorate would vote for party R if there were an election tomorrow. A student at a large college suspects that the proportion of young people who would vote for them is lower. She asks 16 fellow students, chosen at random from the college roll, which party they would vote for. Three choose party R. Show, at the 10% significance level, that this indicates that the reported figure is too high for the young people at the student's college.

The null and alternative hypotheses are $H_0: p = 0.4$ and $H_1: p < 0.4$ respectively.

Let X be the number of students who choose party R. Under H_0, $X \sim B(16, 0.4)$.

$$P(X \leqslant 3) = P(X = 0) + P(X = 1) + P(X = 2) + P(X = 3)$$

$$= \binom{16}{0}(0.4)^0(0.6)^{16} + \binom{16}{1}(0.4)^1(0.6)^{15}$$

$$+ \binom{16}{2}(0.4)^2(0.6)^{14} + \binom{16}{3}(0.4)^3(0.6)^{13}$$

$$= 0.000\,28\ldots + 0.003\,00\ldots + 0.015\,04\ldots + 0.046\,80\ldots$$

$$= 0.065\,14\ldots$$

$$= 6.51\% < 10\%.$$

The result is significant at the 10% level. This indicates that the reported figure is too high for young people at the student's college.

Example 7.1.2
In order to test a coin for bias it is tossed 12 times. The result is 9 heads and 3 tails. Test, at the 10% significance level, whether the coin is biased.

This is a two-tail test since, before the coin is tossed, there is no indication in which direction, if any, it might be biased. If the coin is unbiased, the probability of a head (or a tail) is $p = \frac{1}{2}$.

The null and alternative hypotheses are $H_0: p = \frac{1}{2}$ and $H_1: p \neq \frac{1}{2}$ respectively.

Let X be the number of heads resulting from 12 tosses. Under H_0, $X \sim B(12, 0.5)$. On average you would expect 6 heads. The observed value of 9 is on the high side so,

$$P(X \geqslant 9) = P(X = 9) + P(X = 10) + P(X = 11) + P(X = 12)$$

$$= \binom{12}{9}\left(\frac{1}{2}\right)^9\left(\frac{1}{2}\right)^3 + \binom{12}{10}\left(\frac{1}{2}\right)^{10}\left(\frac{1}{2}\right)^2$$

$$+ \binom{12}{11}\left(\frac{1}{2}\right)^{11}\left(\frac{1}{2}\right)^1 + \binom{12}{12}\left(\frac{1}{2}\right)^{12}\left(\frac{1}{2}\right)^0$$

$$= 0.053\,71\ldots + 0.016\,11\ldots + 0.002\,92\ldots + 0.000\,24\ldots$$

$$= 0.072\,99\ldots$$

$$= 7.30\%, \text{ correct to 3 significant figures.}$$

Since this is a two-tail test at the 10% significance level, this probability must be compared with 5%. Since $7.30\% > 5\%$ the result is not significant and the null hypothesis is not rejected: there is insufficient evidence, at the 10% level, to say that the coin is biased.

> To carry out a hypothesis test on a discrete variable, calculate the probability of the observed or a more extreme value and compare this probability with the significance level. For a one-tail test, reject the null hypothesis if this probability is less than the significance level; for a two-tail test reject the null hypothesis if this probability is less than half of the significance level.

Exercise 7A

1   A large housing estate contains a children's playground, and on one particular evening 12 boys and 6 girls were playing there. Assuming these children are a random sample of all children living on the estate, test, at the 10% significance level, whether there are equal numbers of boys and girls on the estate.

2   An advertisement in a newspaper inserted by a car dealer claimed, 'At least 95% of our customers are satisfied with our services.'

In order to check this statement a random sample of 25 of the dealer's customers were contacted and 22 agreed that they were satisfied with the dealer's services. Carry out a test, at the 5% significance level, of whether the data support the claim.

3   The lengths of nails produced by a machine have a normal distribution with mean 2.5 cm. A random sample of 16 nails is selected from a drum containing a large number of these nails. The nails are measured and 13 are found to have length greater than 2.5 cm. Test, at the $2\frac{1}{2}$% significance level, whether the mean length of the nails in the drum is greater than 2.5 cm. State where in the test the information that the nails have a normal distribution is used.

4   A test of 'telepathy' is devised using cards with faces coloured either red, green, blue or yellow, in equal numbers. When a card is placed face down on a table, at random, Dinesh believes he can forecast the colour of the face correctly. The cards are thoroughly mixed, one is selected and placed face down on the table and Dinesh forecasts the colour of the face. This procedure was repeated 10 times.

It is given that Dinesh was correct on 5 occasions. Test, at the 5% significance level, whether the probability that Dinesh forecasts the colour of a card correctly is greater than $\frac{1}{4}$.

5   A dice is suspected of being loaded to give fewer sixes when thrown than would be expected from a fair dice. In order to test this suspicion, the dice is thrown 30 times. One six is obtained.

(a)   Find the p-value of a test of the suspicion.

(b)   Is the suspicion confirmed at the 5% level?

6   A magazine article reported that 70% of computer owners use the internet regularly. Marie believed that the true figure was different and she consulted 12 of her friends who owned computers. Twelve said that they were regular users of the internet.

(a)   Test Marie's belief at the 10% significance level.

(b)   Comment on the reliability of the test in the light of Marie's sample.

7.2 Testing a population proportion for large samples

When the sample is large, you can calculate probabilities by using the fact that the binomial distribution can be approximated by the normal distribution. The following examples illustrate the method.

Example 7.2.1

In a multiple choice paper a candidate has to select one of four possible answers to each question. On a paper with 100 questions a student gets 34 correct answers. Test, at the 5% significance level, the null hypothesis that the student is guessing the answers.

If the student is guessing the answers then the probability that any one answer is correct is $\frac{1}{4}$. If the student is not guessing then the proportion of correct answers should be greater than this.

The null and alternative hypotheses are $H_0: p = \frac{1}{4}$ and $H_1: p > \frac{1}{4}$ respectively.

Let X be the number of correct answers. Under H_0, $X \sim B\left(100, \frac{1}{4}\right)$.

The probability of the observed or a more extreme value is $P(X \geqslant 34)$. In order to find this probability, $X \sim B\left(100, \frac{1}{4}\right)$ is approximated by a normal distribution with

$$\mu = np = 100 \times \tfrac{1}{4} = 25,$$

$$\sigma^2 = npq = 100 \times \tfrac{1}{4} \times \tfrac{3}{4} = 18.75.$$

So $X \sim B\left(100, \frac{1}{4}\right)$ is approximated by $V \sim N(25, 18.75)$.

$$
\begin{aligned}
P(X \geqslant 34) &\approx P(V \geqslant 33.5) \qquad \text{(using a continuity correction)} \\
&= P\left(Z \geqslant \frac{33.5 - 25}{\sqrt{18.75}}\right) \\
&= P(Z \geqslant 1.963) \\
&= 1 - \Phi(1.963) = 1 - 0.9752 \\
&= 0.0248 = 2.48\% < 5\%.
\end{aligned}
$$

The null hypothesis can be rejected: there is evidence, at the 5% significance level, that the student is not guessing.

Alternatively, there is no need to carry the calculation beyond the point where the value of Z ($= 1.963$) is obtained. Instead you can use the idea of a rejection region, which was developed in Section 6.4. For a one-tail test at the 5% significance level, the rejection region for the test statistic Z is $Z \geqslant 1.645$. The observed value of Z is in the rejection region and H_0 is rejected, as before. You can use whichever method you prefer.

The following is an example of a two-tail test.

Example 7.2.2

If births are equally likely on any day of the week then the proportion of babies born at the week-end should be $\frac{2}{7}$. Out of a random sample of 490 children it was found that 132 were born at the week-end. Does this provide evidence, at the 5% significance level, that the proportion of babies born at the week-end differs from $\frac{2}{7}$?

The null and alternative hypotheses are $H_0 : p = \frac{2}{7}$ and $H_1 : p \neq \frac{2}{7}$ respectively.

Let X be the number of babies born at the week-end. Under H_0, $X \sim B\left(490, \frac{2}{7}\right)$.

Approximate X by $V \sim N\left(490 \times \frac{2}{7}, 490 \times \frac{2}{7} \times \frac{5}{7}\right) = N(140, 100)$.

Since 132 is lower than 140, the mean under H_0, find

$$P(X \leqslant 132) \approx P(V \leqslant 132.5)$$
$$= P\left(Z \leqslant \frac{132.5 - 140}{\sqrt{100}}\right)$$
$$= P(Z \leqslant -0.75).$$

For a two-tail test at the 5% significance level the rejection region is $|Z| \geqslant 1.96$. The calculated value of $|Z|$ is 0.75, which does not lie in the rejection region. The value of $\frac{2}{7}$ for the proportion of babies born at the week-end is accepted.

If you wish to work with probabilities, you would continue the calculation to find
$P(Z \leqslant -0.75) = 0.2266 = 22.66\%$. *Since* $22.66\% > 2\frac{1}{2}\%$ *(remember that this is a two-tail test) the null hypothesis is not rejected.*

Exercise 7B

1   A jar contains a large number of coloured beads, some of which are red. A random sample of 80 of these beads is selected and 19 are found to be red. Test, at the 10% significance level, whether 30% of the beads in the jar are red.

2   A new cold relief drug is tested for effectiveness on 150 volunteers, and 124 of them found the drug beneficial. The manufacturers believe that more than 75% of people suffering from a cold will find the drug beneficial. Test the manufacturers' belief at the $2\frac{1}{2}\%$ significance level.

3   A selling price of $22 000 has been proposed for a new model of car manufactured by a large company. The price will be adopted if more than 40% of potential customers are willing to pay that price. A sample of 50 potential customers were supplied with the car's specification and 29 agreed to pay the proposed price. Carry out a test, at the 1% significance level, to decide whether the company should adopt the price of $22 000.

4   A parcel delivery service claims that at least 80% of their parcels are delivered within 48 hours of posting. A check on 200 parcels found that 152 were delivered within 48 hours of posting. Test the delivery service's claim at the 5% significance level.

5 In Zimbabwe, 8% of males are colour-blind. A random sample of 500 males was selected from a town in Zimbabwe and 53 men were found to be colour-blind. Find the *p*-value of the test of whether the town contains a different proportion of colour-blind males than in Zimbabwe as a whole.

What would be the conclusion of the test at the following significance levels:

(a) 5%, (b) 2%?

6 The drop-out rate of students enrolled at a certain university is reported to be 13.2%. The Dean of Students suspects that the drop-out rate for science students is greater than 13.2%, and she examines the records of a random sample of 95 of these students. The number of drop-outs was found to be 20. Test the Dean's suspicion at the 2% significance level.

7.3 Testing a population mean for a Poisson distribution

The ideas developed in this and the previous chapter can also be applied when the Poisson distribution is a suitable model for the population from which the sample is drawn.

Example 7.3.1

In the past an office photocopier has failed, on average, three times every two weeks. A new, more expensive, photocopier is on trial which the manufacturers claim is more reliable. In the first four weeks of use this new photocopier fails once. Assuming that the failures of the photocopier occur independently and at random, test, at the 5% significance level, whether there is evidence that the new photocopier is more reliable than the old one.

> If the failures of the photocopier occur independently and at random then the number of failures in a given time interval can be modelled by a Poisson distribution. The null hypothesis will be a theory of 'no change'; that is, the new photocopier fails 3 times every two weeks on average. So λ, the mean number of failures per week, is equal to 1.5.
>
> The null and alternative hypotheses are $H_0:\lambda = 1.5$ and $H_1:\lambda < 1.5$ respectively.
>
> Let X be the number of failures in a four-week period. Under H_0, $X \sim \text{Po}(4 \times 1.5) = \text{Po}(6)$.
>
> The observed value of X is 1.
>
> $$P(X \leqslant 1) = e^{-6} + e^{-6}\frac{6}{1!}$$
> $$= 0.002\,478\ldots + 0.014\,872\ldots$$
> $$= 0.017\,35\ldots$$
> $$= 1.74\%, \text{ correct to 3 significant figures.}$$
>
> Since $1.74\% < 5\%$, the result is significant at the 5% significance level and there is evidence that the new photocopier is more reliable than the old one.

Encouraged by this result, the office decides to continue using the new photocopier for the rest of the year. The results are analysed in the following example.

Example 7.3.2

The number of failures in the rest of the first year is 57. Using the results for the whole year, test, at the 5% significance level, whether the new photocopier is more reliable than the old one.

The null and alternative hypotheses are $H_0: \lambda = 1.5$ and $H_1: \lambda < 1.5$ respectively, as before, where λ is the mean number of failures per week.

Let X be the number of failures in 52 weeks. Under H_0, $X \sim Po(52 \times 1.5) \sim Po(78)$.

The observed value of X is $1 + 57$, which is 58. Since $\lambda > 15$, $P(X \leqslant 58)$ can be calculated using the normal approximation (see Section 1.6).

$X \sim Po(78)$ is approximated by $Y \sim N(78, 78)$ with a continuity correction.

$$P(X \leqslant 58) = P(Y < 58.5)$$
$$= P\left(Z < \frac{58.5 - 78}{\sqrt{78}}\right)$$
$$= P(Z < -2.208).$$

For a one-tail test at the 5% significance level the rejection level is $Z \leqslant -1.645$. The calculated value of Z is -2.208. Since this lies in the rejection region the null hypothesis is rejected. The test confirms the result of the previous test (in Example 7.3.1) that the new photocopier is more reliable than the old one.

If you wish to work with probabilities, you would continue the calculation to find $P(Z < -2.208) = 0.0135 = 1.35\%$. *Since $1.35\% < 5\%$, the null hypothesis is rejected.*

7.4 Practical activities

1 A matter of taste   Can people distinguish between different varieties of cola-flavoured drink? Choose two brands. Present each person in your sample with two glasses of one brand and one of the other and explain that you are going to ask them to pick the odd one out. You should toss a coin in order to decide which brand is presented twice. The glasses should be given in a random order and it would be helpful to give each person a glass of water to clear their palate if necessary. Allow people to retaste the drinks if they wish. If there are noticeable differences in colour between the two brands then you could use a mixture of differently coloured plastic beakers in order to conceal this. Ask as large a sample of people as possible and record whether or not they pick the odd drink out correctly.

What is the probability that a person will correctly pick the odd drink out if there is no detectable difference in flavour? Carry out a hypothesis test taking as your null hypothesis that p has this value.

2 Crossed arms   Most people have a very marked preference for the way in which they cross their arms. Try this out for yourself. Cross your arms and note which forearm is on top. Now try to cross your arms with the other forearm on top. You will probably find that this requires a bit of thought! Is a person chosen at random equally likely to prefer left over right as right over left? Ask a number of people to cross their arms and note which forearm is on top. Collect results from as large a sample as possible and test $H_0: p = \frac{1}{2}$, $H_1: p \neq \frac{1}{2}$, where p is the probability that a person crosses their arms right over left.

3 A test of 'telepathy' Carry out an experiment similar to the one described in Question 4 of Exercise 7A. There is no need to limit the experiment to 10 trials.

Exercise 7C

1 The random variable X has a Poisson distribution with mean λ. A single observation of X has the value 4. Test the null hypothesis $\lambda = 2$ against the alternative hypothesis $\lambda > 2$ at the 5% significance level.

2 The number of car accidents that occur along a certain stretch of road may be assumed to have a Poisson distribution with mean 4 per week. In the first two weeks after a new warning sign had been erected, 3 accidents occurred on the road. Test, at the 5% significance level, whether this indicates a reduction in the mean accident rate.

During the next 3 weeks 11 accidents occurred on the road. Does this extra information alter the conclusion of the test?

3 A company manufactures 5 amp fuses and, under normal conditions, 7% of the fuses are faulty. They are packed in boxes of 60.

    (a) Explain why the number of faulty fuses in a randomly chosen box has an approximate Poisson distribution.

    (b) A box randomly chosen from a day's production has 1 faulty fuse. Test, at the 5% significance level, whether the percentage of faulty fuses on that day is lower than 7%.

4 The number of errors in a page of manuscript word-processed by my secretary has a Poisson distribution with mean 1.4. I received a manuscript on a particular day and counted 4 errors on a randomly chosen page. Test, at the 10% significance level, whether this indicates that the manuscript was not word-processed by my secretary.

5 Between January and March, the number of emergency calls received by a power company occur randomly at a uniform rate of 6 per day. During three days in May the power company received a total of 9 emergency calls. Find the p-value of a test of whether the mean number of emergency calls per day, in May, is less than 6.

What can be concluded at the 5% significance level?

6 Over a period of time it has been found that the mean number of letters per week passing through a small sorting office is 3245. In the week following a campaign to promote letter writing 3455 letters passed through the office. Assuming that the number of letters per week can be modelled by a Poisson distribution, test, at the 5% level, whether there is evidence that the publicity campaign has been effective.

<hr>

Miscellaneous exercise 7

Part A contains questions on this chapter only. Part B contains questions on the contents of both Chapters 6 and 7.

Part A

1 A leading newspaper reported that 2 out of every 3 female football club fans were able to explain the offside law correctly. Gaussian Rovers supporters believed that more of their female fans could explain the law correctly. To prove their point a random sample of 20 female Gaussian Rovers fans was questioned outside Bell Park after a match, and 17 were able to explain the offside law correctly.

   (a) Carry out a test, at the 10% significance level, of the hypothesis that the proportion of female Gaussian Rovers fans that can explain the offside law correctly is more than $\frac{2}{3}$.

   (b) Have the supporters 'proved their point'?

2 The probability of a drawing pin landing point up when dropped onto a horizontal floor from a height of one metre is denoted by p. When the drawing pin is dropped 25 times it lands point up 5 times.

   Test the null hypothesis $p = 0.4$ against the alternative hypothesis $p < 0.4$ at the $2\frac{1}{2}\%$ significance level.

3 It is suggested that one-third of all mathematicians are left-handed. In a survey, 51 out of 174 mathematicians were found to be left-handed. Assuming that the sample was random, carry out a test, at the 5% significance level, of whether or not the sample confirms the suggestion.

   (OCR, adapted)

4 In the promotion of Doggo, a new animal food, it was asserted that more dogs prefer the new food to the current brand leader. In a test of this assertion, 40 dogs were given a choice of Doggo and the current brand leader.

   (a) Find the smallest number of dogs that would have to prefer Doggo for the promoter's assertion to be accepted at the 5% significance level.

   (b) What then is the significance level of the test?

5 A supermarket buys a large batch of plastic bags from a manufacturer to be used in the store. In previous batches 7% of the bags were defective. A quality control manager wishes to test whether the batch has a higher defective rate than 7%, in which case the batch will be returned to the manufacturer. He examines 125 randomly selected bags and finds that 14 are defective. Carry out the manager's test at the 3% significance level and state whether he should return the batch.

6 During the period from May 1999 to April 2002, 18 lap-top computers were lost by employees at an international company. After a vigorous enquiry it was hoped that the rate of loss would drop.

   (a) State what must be assumed for the number of lap-top computers lost during a fixed period of time to have a Poisson distribution.

   (b) Find the greatest number of lap-top computers that can be lost during the next year in order to be significant, at the $2\frac{1}{2}\%$ significance level, of a drop in the loss rate.

7 A machine that weaves a carpet of width 2 m produces slight flaws in the carpet at a rate of 1.8 per metre length.

(a) State what must be assumed for the number of flaws in a given length of carpet to have a Poisson distribution.

(b) After the machine is given an overhaul a random sample of 3 m length of the carpet is examined and found to have 2 flaws. Test, at the 5% significance level, whether the rate of incidence of the flaws has decreased.

(c) A further 20 m length of the carpet is found to have 9 flaws. Pooling the two results, determine whether the conclusion of the above test changes.

8 Wild flowers of a certain species grow randomly in a forest area and at a uniform rate of 7.6 per 10 000 m^2.

(a) Suggest a suitable probability distribution of the number of the flowers that grow in an area of 2500 m^2 of the forest.

(b) After an unusually busy month, when the forest was visited by a large number of tourists, the forest managers wished to investigate whether the number of flowers of the species had decreased. In a pilot test a randomly chosen 2500 m^2 of the forest was studied and found to contain no flower of the species. Test whether this indicates, at the 5% significance level, that the number of flowers of the species has decreased.

(c) In a further study, the managers examined 50 randomly selected areas of 2500 m^2 of forest. In 13 of the areas no flower of the species was found. Using a significance level of 5%, test whether this indicates that the number of flowers of the species has decreased.

Part B

9 The mass of Vitamin E in a capsule made by a drug company is normally distributed with mean μ mg and standard deviation 0.058 mg. A random sample of 16 capsules was analysed and the mean mass of Vitamin E was 4.97 mg. Test, at the 2% significance level, the null hypothesis $\mu = 5.00$ against the alternative hypothesis $\mu < 5.00$.

10 The proportion of patients who suffer an allergic reaction to a drug used to treat a particular medical condition is assumed to be 0.036. When 500 patients were treated with the drug, 28 suffered an allergic reaction. Test, at the 5% significance level, whether the quoted figure of 0.036 is an underestimate.   (OCR, adapted)

11 Students using a college canteen for lunch paid an average of £1.74 during 2000. A new catering company was appointed in 2001 and after the company had provided food for one month a random sample of 60 students was asked to state the amount spent daily on lunch. The results, £ x, were summarised by $\sum x = 118.80$ and $\sum x^2 = 271.81$. Test, at the 3% significance level, whether the average amount paid by students for lunch had increased.

The person who carried out the above test included in his report the following incorrect statement. Give a corrected version. 'It is not necessary for the population to have a normal distribution since the sample size is large and the Central Limit Theorem states that any sufficiently large sample is normal.'   (OCR, adapted)

12 A firm manufactures glass vases and the proportion of defective vases is 0.2. The quality control department wants to reduce the proportion of defective vases and makes changes to the manufacturing procedure. A random sample of 25 vases is then examined and 2 are found to be defective. Find the p-value of a test of whether the proportion of defective vases has been reduced. Give the conclusion of the test at the 10% significance level.

13 In a certain town, the radioactive count due to background radiation is, on average, 6 particles per minute. Following an accident at a nearby nuclear power station, a count of 72 particles was obtained for a time interval of 10 minutes. Test, at the 5% level, whether this result indicates that the average background radiation has increased.

What count would have to be obtained for a time interval of 10 minutes in order for the result to be significant at the 5% level?

14 The age, x years, of each of the women giving birth at a maternity hospital in the year 2000 was recorded correct to 1 decimal place. In a random sample of 120 cases it was found that $\sum x = 3468$ and $\sum x^2 = 110\,151$. The ages of all women giving birth in the hospital has mean μ years and standard deviation σ years.

(a) Calculate unbiased estimates of μ and σ^2.

(b) Carry out a significance test, at the 5% significance level, in which the null hypothesis $\mu = 30.0$ is tested against the alternative hypothesis $\mu < 30.0$.

(c) It was subsequently found that an error had been made in recording the value of $\sum x^2$ for the sample. Find the greatest actual value of $\sum x^2$ for the sample which would give rise to a different conclusion in part (b).   (OCR, adapted)

15 On average, 4 out of 5 new television sets of a particular brand are fault-free during the first year of purchase. A new design is marketed and a random sample of 20 sets is monitored by the manufacturer over a period of a year. The number of fault-free sets during this period was 19.

(a) Test, at the 5% significance level, whether the proportion of fault-free sets of this new design is greater than 4 out of 5.

(b) What is the smallest significance level at which it would be accepted that the proportion of fault-free sets is greater than 4 out of 5?

16 The lifetime of a Brightray battery has a normal distribution with mean μ hours and standard deviation $4\frac{3}{4}$ hours. A random sample of 36 batteries is selected and the sum of the lifetimes is found to be 585 hours.

(a) Show that the null hypothesis $\mu = 15\frac{1}{2}$ cannot be rejected, at any significance level, in favour of the alternative hypothesis $\mu < 15\frac{1}{2}$.

(b) Find the p-value of the test of the null hypothesis $\mu = 15\frac{1}{2}$ against the alternative hypothesis $\mu > 15\frac{1}{2}$.

What is the conclusion of the test at the 5% significance level?

17 The average number of calls received each day by a telephone Help Line was 1.5. After a publicity campaign in the press and on radio, it was found the total number of calls to the Line, over a period of 2 days, was 5.

(a) State a suitable probability distribution to use in a test of whether the daily average number of calls to the Line has increased.

What must be assumed for the validity of the chosen probability distribution?

(b) Carry out the test at the 5% significance level.

18 Occasionally the intruder alarm at a warehouse goes off incorrectly. On average, such false alarms occur 6 times a year. During a particular month there were three false alarms. Assuming that false alarms can be treated as random events, at what significance level can it be accepted that the mean number of false alarms for this period is greater than 6 per year?

19 A student answers a test consisting of 12 multiple choice questions, in each of which the correct response has to be selected from four possible given answers. The student only gets 2 of the questions correct and the teacher claims that 'this shows that the student did worse than anyone would do just by guessing'. Denoting the probability of the student answering a question correctly by p, carry out a suitable test to investigate the teacher's claim at the 10% significance level.

Hence state with a reason whether you agree with the teacher's claim.   (OCR, adapted)

20 Metal washers are produced in a factory in batches of 10 000. In the production process occasional faults occur at random, resulting in defective washers being produced. On average, the proportion of defective washers is 0.03%. Explain why you would expect a Poisson distribution to provide a suitable model for the number of defective washers in a batch.

Using a Poisson distribution, show that approximately 5% of batches contain no defective washers.

Following a change in the manufacturing procedure it is suspected that the proportion of batches containing no defective washers is now greater that 5%. It is desired to test whether this is the case. State appropriate hypotheses for a significance test.

Carry out the test, at the 10% significance level, given that in a random sample of 10 batches there were 2 containing no defective washers.   (OCR, adapted)

21 Cartons of milk are tested by a consumer association both for quantity and for ease of opening. A random sample of 100 cartons is examined and the quantity, x litres, of milk in each carton is determined. The results are summarised by $\sum(x-1) = 1.21$ and $\sum(x-1)^2 = 0.5377$. Test, at the 5% significance level, whether the mean quantity of milk in a carton is 1.005 litres against the alternative hypothesis that it is greater than 1.005 litres.

A cheaper design of carton is introduced and the consumer association decides to carry out a new test. A random sample of 100 cartons is tested and 53 are found easy to open. Test, at the 5% significance level, whether the proportion of cartons that are easy to open is less than 65%.   (OCR, adapted)

22   A store discovers that its credit card machine rejects, on average, one card in every 890 transactions. Let X denote the number of rejections in a randomly chosen 2136 transactions.

   (a) Explain why the distribution of X may be approximated by a Poisson distribution.

   (b) On a particular day when there were 2136 transactions the number of rejected cards was 6. Test, at the 5% significance level, whether there is evidence that the average number of rejected cards has increased.

23   The manufacturers of a certain type of laser printer state that when used under 'typical business conditions' the toner cartridge in the printer should last for 'approximately 3000 pages'. A survey involving 150 randomly chosen printers used in business was carried out and the number of pages, x, printed before a new toner cartridge had to be replaced is summarised by $\sum x = 4.731 \times 10^5$ and $\sum x^2 = 1.620 \times 10^9$, each correct to 4 significant figures. Carry out a test, at the $2\frac{1}{2}\%$ significance level, that the mean number of pages that can be printed is 3000 against the alternative hypothesis that it is greater than 3000.

The printer manufacturers consider that their printer produces noticeably higher quality print than other, comparable machines. Of a random sample of 150 users who were asked their opinions about this, 51 agreed with the manufacturers and the remainder either disagreed or had no opinion. Test, at the 5% significance level, the null hypothesis that the proportion of users agreeing with the printer manufacturers about the print quality is 40% against the alternative hypothesis that it is not 40%.   (OCR, adapted)

8 Errors in hypothesis testing

This chapter investigates the situation where the wrong conclusion is drawn from a hypothesis test. When you have completed it you should

- know what Type I and Type II errors are
- be able to calculate probabilities of Type I and Type II errors in the context of the normal, binomial and Poisson distributions.

8.1 Type I and Type II errors

When you carry out a hypothesis test your final step is to reject or to accept the null hypothesis. For example, in the situation described in Section 7.3, where a new photocopier was being tested, the users of the photocopier had to choose between the conclusions

    (a) the new photocopier is better than the old one or
    (b) the new photocopier is not better than the old one.

The result of the hypothesis test will help the users to decide on their next action. If they came to conclusion (a) they would probably decide to keep the new photocopier; if they came to conclusion (b) they would probably keep the old photocopier. Similarly, the teachers described in Section 6.1, who were trying out a new reading scheme, had to choose between

    (c) the new reading scheme is better than the old one or
    (d) the new reading scheme is not better than the old one.

If they came to conclusion (c) they would probably introduce the new scheme; if they came to conclusion (d) they would probably stick to their current reading scheme.

When such a decision is made after carrying out a hypothesis test, it may be either correct or incorrect. You can never be absolutely certain that you have made the right decision because you have to rely on a limited amount of evidence. For example, the photocopier can only be tested for a limited period; the reading scheme can only be tested on a sample of children. The situation is similar to that in a trial where the defendant is found either guilty or not guilty on the basis of the evidence brought forward. In this case there are four possible situations (which are mutually exclusive).

The defendant is innocent and is found not guilty: in this case the decision is correct.
The defendant is innocent but is found guilty: in this case the decision is incorrect.
The defendant is guilty but is found not guilty: in this case the decision is incorrect.
The defendant is guilty and is found guilty: in this case the decision is correct.

Suppose that in a criminal court of law a defendant is assumed innocent unless found guilty 'beyond reasonable doubt'. The initial assumption of innocence is equivalent to the null hypothesis and the theory that the defendant is guilty is equivalent to the alternative hypothesis. Deciding what constitutes 'beyond reasonable doubt' is equivalent to setting a significance level. Similarly in a hypothesis test there are four possible situations, again mutually exclusive.

> H_0 is true and H_0 is accepted: in this case the decision is correct.
> H_0 is true but H_0 is rejected: in this case the decision is incorrect.
> H_0 is not true but H_0 is accepted: in this case the decision is incorrect.
> H_0 is not true and H_0 is rejected: in this case the decision is correct.

You can see that there are two different ways in which an incorrect decision could be made. In order to distinguish between them they are called Type I and Type II errors.

> A **Type I error** is made when a true null hypothesis is rejected.
>
> A **Type II error** is made when a false null hypothesis is accepted.

Making an incorrect decision can be costly in various ways. For example, suppose that a fire alarm was tested to see whether it was still functioning correctly after a power cut. You might take as the null and alternative hypotheses

> H_0: the alarm is functioning correctly,
> H_1: the alarm is not functioning correctly.

A Type II error in this situation would mean that you assumed that the alarm was functioning correctly when in fact it was not. This could result in injury, loss of life or damage to property. A Type I error would mean that you thought the alarm was not working correctly when in fact it was. This could mean expenditure on unnecessary repairs or replacement.

Try to analyse in a similar way the 'costs' of making Type I and Type II errors for
(a) the photocopier example ,
(b) the reading scheme example.

The examples given in this chapter should make you appreciate that it is important to assess the risk of making errors when carrying out a hypothesis test. In order to do this you have to calculate

$$P(\text{Type I error}) = P(\text{rejecting } H_0 \mid H_0 \text{ true})$$
and $$P(\text{Type II error}) = P(\text{accepting } H_0 \mid H_0 \text{ false}).$$

The following sections show you how these probabilities are calculated for the different types of test which you have met in Chapters 6 and 7.

8.2   Type I and Type II errors for tests involving the normal distribution

If you look back to Section 6.3, which considered continuous variables, you will see that the probability of the test statistic falling in the rejection region, when H_0 is true, is equal to the significance level of the test. If the test statistic falls in the rejection region then H_0 will be rejected when it is in fact true; that is, a Type I error will be made.

> When the distribution of the test statistic is continuous,
> P(Type I error) is equal to the significance level of the test.

The choice of significance level for a hypothesis test is thus related to the value of P(Type I error) which you are prepared to accept. The choice of a significance level should depend in the first instance on how serious the consequences of a Type I error are. The more serious the consequences, the lower the value of the significance level which should be used. For example, if the consequences of a Type I error were not serious, you might use a significance level of 10%; if the consequences were very serious you might use a significance level of 0.1%.

Consider the reading scheme example. A Type I error in this case would mean that the new scheme was adopted even though it did not produce better results. As a result money would be wasted on a new scheme which was no better than the old. If the new scheme is not any better and the teachers use a 5% significance level then there a 1 in 20 chance that the money will be wasted. If the new material is very costly then the teachers might feel that such a risk is unacceptable and choose a significance level of 1% or even less depending on the resources of their school. If on the other hand the new scheme is not very expensive, or they need to replace their reading material anyway, then they might take a significance level of 10% or even 20%.

A Type II error involves accepting a false null hypothesis, which means that you fail to detect a difference in μ. You would expect the probability of this happening to depend on how much μ has changed: if there is a small difference in μ it could easily go undetected but if there is a big difference in μ then you would expect to detect it. This is why the alternative hypothesis has to be defined more exactly before P(Type II error) can be calculated.

The following example illustrates the method.

Example 8.2.1
A machine fills 'one litre' water bottles. When the machine is working correctly the contents of the bottles are normally distributed with mean 1.002 litres and standard deviation 0.002 litres. The performance of the machine is tested at regular intervals by taking a sample of 9 bottles and calculating their mean content. If this mean content falls below a certain value, it is assumed that the machine is not performing correctly and it is stopped.

(a) Set up null and alternative hypotheses for a test of whether the machine is working correctly.
(b) For a test at the 5% significance level, find the rejection region taking the sample mean as the test statistic.
(c) Give the value for the probability of a Type I error.
(d) Find P(Type II error) if the mean content of the bottles has fallen to the nominal value of 1.000 litre.
(e) Find the range of values of μ for which the probability of making a Type II error is less than 0.001.

(a) $H_0: \mu = 1.002$ (the machine is working correctly);

$H_1: \mu < 1.002$ (the mean content has fallen).

(b) Under H_0, $\overline{X} \sim N\left(1.002, \dfrac{0.002^2}{9}\right)$.

For a one-tail test for a decrease at the 5% level, the rejection region for the test
statistic Z is $Z \leqslant -1.645$. Since $Z = \dfrac{\overline{X} - \mu}{\sqrt{\dfrac{\sigma^2}{n}}}$ this means that $\dfrac{\overline{X} - 1.002}{\sqrt{\dfrac{0.002^2}{9}}} \leqslant -1.645$.

Rearranging gives the rejection region for the sample mean as
$\overline{X} \leqslant 1.000\,90\ldots = 1.0009$, correct to 4 decimal places.

(c) For a continuous test statistic, $P(\text{Type I error}) = \text{significance level} = 0.05$.

(d) $P(\text{Type II error}) = P(\text{accepting } H_0 \mid H_0 \text{ false})$

$\qquad\qquad = P(\overline{X} > 1.0009 \mid \mu = 1.000)$,

that is $P(\overline{X}$ is in acceptance region $\mid \mu$ is no longer 1.002 but 1.000$)$.

$$P(\overline{X} > 1.0009 \mid \mu = 1.000) = P\left(Z > \dfrac{1.000\,90\ldots - 1.000}{\sqrt{\dfrac{0.002^2}{9}}} \right) = P(Z > 1.355)$$

$$= 1 - \Phi(1.355) = 1 - 0.9123 = 0.0877.$$

These results are illustrated in Fig. 8.1. The broken curve shows the distribution of $\overline{X}$
if H_0 is true, and the solid curve shows the distribution if H_1 is true and the mean
has fallen to 1.000. The hatched area shows $P(\text{Type I error})$ and the solid shaded
area shows $P(\text{Type II error})$.

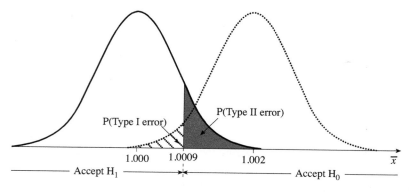

Fig. 8.1. Type I and Type II errors.

(e) First find the value of z for which the probability of making a Type II error is
0.001. Looking back to part (d) this would require $\Phi(z) = 0.999$ and hence $z = 3.090$.

The value of μ which gives this value can be obtained from

$$z = 3.090 = \frac{1.0009 - \mu}{\sqrt{\dfrac{0.002^2}{9}}}.$$

Solving gives $\mu = 0.9988$.

So for $\mu < 0.9988$ the probability of making a Type II error is less than 0.001.

In Example 8.2.1 the consequence of setting the significance level at 5% is that there is a probability of 5% of stopping the machine unnecessarily when it is working correctly. With this significance level the probability of failing to detect that the mean content of the bottles has fallen to the nominal value of 1.000 litre is 8.77%. It is interesting to see what happens to $P(\text{Type II error})$ when a lower significance level is used. This is done in the next example.

Example 8.2.2

Repeat Example 8.2.1 parts (b) to (d) with a significance level of 1%.

(b) For a one-tail test for a decrease at the 1% level, the rejection region for the test statistic Z is $Z \leqslant -2.326$ so

$$\frac{\overline{X} - 1.002}{\sqrt{\dfrac{0.002^2}{9}}} \leqslant -2.326.$$

Rearranging gives the rejection region for the sample mean as $\overline{X} \leqslant 1.000\ 45$.

(c) $P(\text{Type I error}) = \text{significance level} = 0.01$.

(d) $P(\text{Type II error}) = P(\text{accepting } H_0 \mid H_0 \text{ false})$
$$= P(\overline{X} > 1.000\ 45 \mid \mu = 1.000),$$

that is $P(\overline{X}$ is in acceptance region $\mid \mu$ is no longer 1.002 but 1.000$)$.

$$P(\overline{X} > 1.000\ 45 \mid \mu = 1.000) = P\left(Z > \frac{1.000\ 45 - 1.000}{\sqrt{\dfrac{0.002^2}{9}}} \right)$$
$$= P(Z > 0.674)$$
$$= 1 - \Phi(0.674) = 1 - 0.7499 = 0.2501.$$

For this second calculation, at the 1% significance level, the value of $P(\text{Type I error})$ has been reduced but the value of $P(\text{Type II error})$ has increased. You would expect this result from looking at Fig. 8.1. If the critical value is altered so that one type of error increases, the other will decrease. This means that in setting a significance level it may be necessary to assess the risks involved in committing both types of error and balance one against the other. The only way in which both types of error can be reduced at the same time is by taking a larger sample, so that the overlap of the distributions in Fig. 8.1 is reduced.

The following example shows how to calculate P(Type II error) for a two-tail test.

Example 8.2.3
Boxes of dried haricot beans have contents whose masses are normally distributed with mean μ and standard deviation 15 grams. A test of the null hypothesis $\mu = 375$ against the alternative hypothesis $\mu \neq 375$ is carried out at the 5% significance level using a random sample of 16 boxes.

(a) For what values of the sample mean is the alternative hypothesis accepted?
(b) Given that the actual value of μ is 380, find the probability of making a Type II error.   (OCR, adapted)

(a) Under H_0, $\overline{X} \sim N\left(375, \dfrac{15^2}{16}\right)$.

For a two-tail test at the 5% significance level, the rejection region is $|Z| \geqslant 1.96$.

Now $Z = \dfrac{\overline{X} - 375}{\sqrt{\dfrac{15^2}{16}}}$.

You can check that when $Z = 1.96$, $\overline{X} = 382.35$, and when $Z = -1.96$, $\overline{X} = 367.65$. Thus the alternative hypothesis is accepted when $\overline{X} \geqslant 382.35$ or $\overline{X} \leqslant 367.65$.

(b) P(Type II error) $= P\left(382.35 > \overline{X} > 367.65 \mid \mu = 380\right)$

$$= P\left(\frac{382.35 - 380}{\sqrt{\dfrac{15^2}{16}}} > Z > \frac{367.65 - 380}{\sqrt{\dfrac{15^2}{16}}}\right)$$

$$= P(0.627 > Z > -3.293)$$
$$= \Phi(0.627) - (1 - \Phi(3.293))$$
$$= 0.7347 - (1 - 0.9995)$$
$$= 0.734, \text{ correct to 3 decimal places.}$$

Exercise 8A

1   The random variable X has a normal distribution with mean μ and variance 12.8. A test, at the 5% significance level, of the null hypothesis $\mu = 5$ against the alternative hypothesis $\mu > 5$ is carried out using a random sample of 20 observations of X.

(a) Give the rejection region of the test, in terms of the sample mean, $\overline{X}$.

(b) Find the probability of a Type II error in the test when the true value of μ is 7.

2 In a test of the quality of Luxiglow paint, which is intended to cover an area of at least $10\,\text{m}^2$ per litre can, a random sample of 15 cans is tested. The mean area per can covered by the 15 cans is denoted by $\overline{X}\,\text{m}^2$. It may be assumed that the area covered by a can has a normal distribution with standard deviation $0.51\,\text{m}^2$.

(a) Find, in terms of $\overline{X}$, the rejection region of a test, at the $2\frac{1}{2}\%$ significance level, that the mean area covered by all litre cans of the paint is at least $10\,\text{m}^2$.

(b) For a particular sample, $\overline{X} = 10.3$. State the type of error that could not occur.

(c) Given that the mean cover per can of paint is actually $9.6\,\text{m}^2$, calculate the probability of making a Type II error in the test.

3 In a quality control check 5 randomly selected packs of butter are weighed. The masses of all packs of butter may be assumed to have a normal distribution with mean μ grams and standard deviation 2.7 grams. A test of the null hypothesis $\mu = 247$ against the alternative hypothesis $\mu \neq 247$ is carried out at the $\alpha\%$ significance level. It is decided to accept the null hypothesis if the sample mean lies between 245 grams and 249 grams.

(a) Find the value of α.

(b) Given that the actual value of μ is 250, find the probability of making a Type II error in the test.

(c) What can be said about the probability of making a Type II error when the value of μ is greater than 250?

4 The number of daily absences by employees of a large company has mean 1.94 and standard deviation 0.22. A new system of working is introduced in the hope that this will reduce the number of absences, and it is found that there were 68 absences during the first 40 days of the new system. Treating the 40 days as a random sample

(a) test, at the 5% significance level, whether the new system had the desired effect,

(b) calculate the probability of making a Type II error in the test in part (a) when the mean number of absences is actually 1.8,

(c) state, in the context of the question, what is meant by a Type II error.

5 Studies have shown that the time taken for adults to memorise a list of 12 words has mean 3.8 minutes and standard deviation 1.8 minutes. Taking a course in memory techniques is believed to reduce the mean. To investigate this belief a test, at the 5% significance level, is proposed based on a random sample of 36 people who took the course. Each was given the same list of 12 words to memorise.

(a) Find the rejection region of the test in terms of the sample mean time. Assume that the standard deviation remains at 1.8 minutes.

(b) Given that the actual mean time for the 12 words is 2.9 minutes after taking the course, find the probability of making a Type II error in the test.

Suppose now that the test is based on a random sample of 40 people.

(c) Show that the probability of making a Type II error (when the actual mean time is 2.9 minutes) is smaller than that found in part (b).

6 The breaking strength of lengths of wire required in the manufacture of a certain piece of machinery has a normal distribution with mean $30\,\text{N}$ and standard deviation $0.38\,\text{N}$. A random sample of 9 lengths of the wire is tested to determine whether the population mean breaking strength is less than $30\,\text{N}$. A Type I error for the test has probability 0.04.

(a) Find the set of values of the sample mean breaking strength for which it would be accepted that the mean breaking strength is not less than $30\,\text{N}$.

(b) Given that the probability of making a Type II error in the test is to be less than 0.025, find the set of possible values of the actual mean breaking strength.

8.3 Type I and Type II errors for tests involving the binomial distribution

In Section 7.1 you met the idea that, for a discrete distribution, it is not usually possible to find a rejection region which corresponds exactly to the specified (nominal) significance level. You may find it helpful to look back at Section 7.1 before going on to the following example.

Example 8.3.1

An investigator suspects that operatives using a spring balance are reluctant to give 0 as the last value of a recorded weight, for example, 4.10 or 0.30. In order to test her theory she takes a random sample of 40 recorded weights and counts the number, X, which end in 0.

(a) State suitable hypotheses, involving a probability, for a hypothesis test which could indicate whether the operatives avoid ending a recorded weight with 0.

(b) Show that, for a test at the 10% significance level, the null hypothesis will be rejected if $X = 1$ but not if $X = 2$.

(c) State the rejection region for the test in terms of X.

(d) Calculate the value of $P(\text{Type I error})$.

(e) The nominal significance level of this test is 10%. What is the actual significance level of the test?

(a) If operatives do not avoid ending a recorded weight with 0 then the probability that a value chosen at random ends in 0 is 0.1; if they do avoid 0 then this probability will be less than 0.1, so take $H_0: p = 0.1$ and $H_1: p < 0.1$.

(b) Under H_0, $X \sim B(40,0.1)$

$$P(X \leqslant 1) = \binom{40}{0}0.1^0 0.9^{40} + \binom{40}{1}0.1^1 0.9^{39} = 0.014\,78\ldots + 0.065\,69\ldots$$
$$= 0.080\,47\ldots.$$

As $0.080\,47\ldots < 0.1$, H_0 is rejected if $X = 1$.

$$P(X \leqslant 2) = 0.080\,47\ldots + \binom{40}{2}0.1^2 0.9^{38} = 0.222\,80\ldots.$$

As $0.222\,80\ldots > 0.1$, H_0 is not rejected if $X = 2$.

(c) The rejection region is $X \leqslant 1$.

(d) $P(\text{Type I error}) = P(X \leqslant 1 \mid p = 0.1)$. From part (b),

$$P(\text{Type I error}) = P(X \leqslant 1 \mid p = 0.1)$$
$$= 0.080\ 47\ldots = 0.0805, \text{correct to 3 significant figures.}$$

(e) The actual significance level of the test is equal to $P(\text{Type I error})$, that is 8.05%.

This is not quite the same as the desired significance level (10% in this example) and this will often be the case in tests involving discrete variables.

> For a hypothesis test involving a discrete variable the rejection region is defined so that
>
> $$P(\text{test statistic falls in rejection region} \mid H_0 \text{ true})$$
> $$\leqslant \text{nominal significance of the test.}$$
>
> Actual significance level of the test
> $$= P(\text{test statistic falls in rejection region} \mid H_0 \text{ true})$$
> and this is also the probability of a Type I error.

The following example shows how to calculate $P(\text{Type II error})$.

Example 8.3.2

A supplier of orchid seeds claims that their germination rate is 0.95. A purchaser of the seeds suspects that the germination rate is lower than this. In order to test this claim the purchaser plants 20 seeds in similar conditions, counts the number, X, which germinate. He rejects the claim if $X \leqslant 17$.

(a) Formulate suitable null and alternative hypotheses to test the seed supplier's claim.
(b) What is the probability of a Type I error using this test?
(c) Calculate $P(\text{Type II error})$ if the probability that a seed germinates is in fact 0.80.

(a) $H_0 : p = 0.95, \quad H_1 : p < 0.95$.

(b) Under H_0, $X \sim B(20, 0.95)$.

$$P(\text{Type I error}) = P(X \leqslant 17 \mid p = 0.95)$$
$$= 1 - P(X \geqslant 18 \mid p = 0.95)$$
$$= 1 - \binom{20}{18} 0.95^{18} 0.05^2 - \binom{20}{19} 0.95^{19} 0.05^1 - \binom{20}{20} 0.95^{20} 0.05^0$$
$$= 1 - 0.188\ 67\ldots - 0.377\ 35\ldots - 0.358\ 48\ldots$$
$$= 0.075\ 48$$
$$= 0.0755, \text{correct to 3 significant figures.}$$

(c) $P(\text{Type II error}) = P(X \geqslant 18 \mid p = 0.8)$

$$= \binom{20}{18}0.8^{18}0.2^2 + \binom{20}{19}0.8^{19}0.2^1 + \binom{20}{20}0.8^{20}0.2^0$$

$$= 0.136\,90\ldots + 0.057\,46\ldots + 0.011\,52$$

$$= 0.206\,08$$

$$= 0.2061, \text{ correct to 3 significant figures.}$$

The value of P(Type II error) in Example 8.3.2 indicates that the probability that the hypothesis test will fail to detect a fall in the germination rate from 0.95 to 0.80 is not negligible.

For large samples it becomes easier to find a rejection region which gives P(Type I error) close to the required significance level. As you saw in Section 7.2, testing a population proportion for a large sample is done using the normal approximation to the binomial distribution. The following example shows how to calculate P(Type I error) and P(Type II error) in this situation.

Example 8.3.3

A manufacturer claims that the probability that an electric fuse is faulty is no more than 0.03. A purchaser tests this claim by testing a box of 500 fuses. A significance test is carried out at the 5% level using X, the number of faulty fuses in a box of 500, as the test statistic.

(a) For what values of X would you conclude that the probability that a fuse is faulty is greater than 0.03?
(b) Estimate $P(\text{Type I error})$ for this test.
(c) For this test estimate $P(\text{Type II error})$ if the probability that a fuse is faulty is, in fact, 0.06.

(a) The null and alternative hypotheses are $H_0: p = 0.03$, $H_1: p > 0.03$.

Under H_0, $X \sim B(500, 0.03)$.

$$\mu = np = 500 \times 0.03 = 15 \quad \text{and} \quad \sigma^2 = npq = 500 \times 0.03 \times 0.97 = 14.55.$$

Using the normal approximation,

$X \sim B(500, 0.03)$ is approximated by $V \sim N(15, 14.55)$.

For a one-tail test for an increase at the 5% significance level, H_0 is rejected if $Z \geqslant 1.645$. The corresponding rejection region for X is given by

$$\frac{X - 0.5 - 15}{\sqrt{14.55}} \geqslant 1.645 \qquad \text{(where a continuity correction has been included),}$$

giving $X \geqslant 21.77\ldots\ .$

The rejection region is taken as $X \geqslant 22$, by rounding up to the next integer.

(b) $P(\text{Type I error}) = P(X \geqslant 22)$, assuming that H_0 is true.

Using the distribution $V \sim N(15, 14.55)$ as an approximation, $P(X \geqslant 22) = P(V \geqslant 21.5)$, including a continuity correction.

$$P(\text{Type I error}) = P\left(Z \geqslant \frac{21.5 - 15}{\sqrt{14.55}}\right) = P(Z \geqslant 1.704)$$
$$= 1 - 0.9558$$
$$= 0.0442.$$

In this case the actual significance level, 4.42%, is quite close to the nominal significance level of 5%.

(c) If $p = 0.06$, then the distribution of X can be approximated by $V \sim N(500 \times 0.06, 500 \times 0.06 \times 0.94) = N(30, 28.2)$.

$$P(\text{Type II error}) = P(X < 22) = P(V < 21.5)$$
$$= P\left(Z < \frac{21.5 - 30}{\sqrt{28.2}}\right) = P(Z < -1.601)$$
$$= 1 - \Phi(1.601)$$
$$= 1 - 0.9453$$
$$= 0.0547.$$

Exercise 8B

1 A newspaper reported that 55% of households own more than one television set. Each of a random sample of 12 households in a town is contacted and the number of households owning more than one television set is denoted by N. A test of whether the proportion p of households in the town owning more than one television set is greater than 55% is carried out. It is decided to accept that p is greater than 55% if $N > 9$.

(a) Calculate $P(\text{Type I error})$.

(b) Calculate $P(\text{Type II error})$ when the actual value of p is 60%.

2 It is suspected that the dice used in a board game is biased away from a six. In order to test this theory, the dice is rolled 30 times and the number, X, of sixes is counted. If the number of sixes is less than 3 it is accepted that the dice is biased away from a six.

(a) Set up suitable null and alternative hypotheses for testing the theory that the dice is biased away from a six.

(b) Calculate the significance level of the test.

(c) State the value of the probability of a Type I error.

(d) If, in fact, the probability of getting a six with the dice is 0.1, calculate the probability of a Type II error.

3 A drug for treating phlebitis has proved effective in 75% of cases when it has been used. A new drug has been developed which, it is believed, will be more successful and it is used on a sample of 16 patients with phlebitis. A test is carried out to determine whether the new drug has a greater success rate than 75% and the test statistic is X, the number of patients cured by the new drug. It is decided to accept that the new drug is more effective if $X > 14$.

(a) Find α, the probability of making a Type I error.

(b) Find β, the probability of making a Type II error when the actual success rate is 80%.

What can be said about the values of α and β if, with the same decision procedure $(X > 14)$, the sample size was between 17 and 19 inclusive?

4 Of a certain make of electric toaster, 10% have to be returned for service within three months of purchase. A modification to the toaster is made in the hope that it will be more reliable. Out of 24 modified toasters sold in a store none was returned for service within three months of purchase. The proportion of all the modified toasters that are returned for service within three months of purchase is denoted by p.

(a) State, in terms of p, suitable hypotheses for a test.

(b) Test whether there is evidence, at a nominal 10% significance level, that the modified toaster is more reliable than the previous model in that it requires less service.

(c) What is the probability of making a Type I error in the test?

(d) Find the set of values of p for which $P(\text{Type II error}) < 0.25$.

5 A bus company has agreed to supply a new service to a particular area if more than 70% of the people in the area will use the service. A random sample of 120 people living in the area is questioned. Let the number who say they will use the service be denoted by S. The bus company carries out a test at a nominal 2% significance level to decide whether they should run the service. They assume that the responses are truthful.

(a) Find, in terms of S, the rejection region of the test.

(b) Estimate the true significance level of the test.

(c) Let the proportion of all the people in the area who will use the service be denoted by p. Find the approximate value of p_0 such that, for $p > p_0$, $P(\text{Type II error}) < \frac{1}{2}$.

6 It is known that many crimes are committed by people with backgrounds of drug abuse. A proportion of 60% has been suggested and, to investigate this, a researcher undertakes a study of 100 criminals and will carry out a test at a nominal 10% significance level. The null hypothesis is that the proportion of such criminals is 60% and the alternative hypothesis is that the proportion differs from 60%.

(a) Find the rejection region of the test.

(b) Find $P(\text{Type I error})$ for the test.

(c) What is the conclusion of the test if 62 of the 100 criminals have backgrounds of drug abuse?

(d) Find $P(\text{Type II error})$ for the test when the actual proportion is 40%.

7 In the population of the UK, about 8% of females have red hair. A statistics student visiting a city in the UK wishes to test whether the percentage in this city differs from 8%. He plans a test at a nominal 1% significance level and observes a random sample of 500 females in several areas of the city.

(a) Find the actual significance level α of the test.

(b) Find the probability β of making a Type II error in the test when the actual proportion of red-haired females in the city is 4%.

(c) Find the value of β if the sample size were 1000. What general result does this indicate?

8.4 Type I and Type II errors for tests involving the Poisson distribution

Examples involving the Poisson distribution are handled in a similar way to those involving the binomial distribution.

Example 8.4.1

In an intensive survey of dune land it was found that the average number of plants of a particular species was 10 per m^2. After a very dry season it is suspected that these plants have tended to die off. In order to test this hypothesis a randomly chosen area of $1\ m^2$ is selected and the number of these plants, X, growing in it is counted. If X is greater than 4, it is assumed that the weather has no effect. You may assume that the distribution of X can be modelled by a Poisson distribution.

(a) State suitable null and alternative hypotheses.   (b) What is P(Type I error)?
(c) Calculate P(Type II error) if the mean number of plants per square metre has changed to   (i) 8   (ii) 4.

Do you think that the assumption of a Poisson model is likely to be justified in this situation?

(a) $H_0: \lambda = 10$, $H_1: \lambda < 10$.

(b) $P(\text{Type I error}) = P(X \leq 4 \mid \lambda = 10)$

$$= e^{-10} + e^{-10}\frac{10^1}{1!} + e^{-10}\frac{10^2}{2!} + e^{-10}\frac{10^3}{3!} + e^{-10}\frac{10^4}{4!}$$

$$= 0.000\ 045\ldots + 0.000\ 453\ldots + 0.002\ 269\ldots$$
$$0.007\ 566\ldots + 0.018\ 916\ldots$$

$$= 0.02925\ldots$$

$$= 0.0293, \text{correct to 3 significant figures.}$$

(c) (i) $P(\text{Type II error}) = P(X > 4 \mid \lambda = 8) = 1 - P(X \leq 4 \mid \lambda = 8)$

$$= 1 - e^{-8} - e^{-8}\frac{8^1}{1!} - e^{-8}\frac{8^2}{2!} - e^{-8}\frac{8^3}{3!} - e^{-8}\frac{8^4}{4!}$$

$$= 1 - 0.000\ 335\ldots - 0.002\ 683\ldots - 0.010\ 734\ldots$$
$$- 0.028\ 626\ldots - 0.057\ 252\ldots$$

$$= 0.9003\ldots$$

$$= 0.900, \text{correct to 3 significant figures.}$$

(c) (ii) $P(\text{Type II error}) = P(X > 4 \mid \lambda = 4) = 1 \pm P(X \leqslant 4 \mid \lambda = 4)$

$$= 1 - e^{-4} - e^{-4}\frac{4^1}{1!} - e^{-4}\frac{4^2}{2!} - e^{-4}\frac{4^3}{3!} - e^{-4}\frac{4^4}{4!}$$

$$= 1 - 0.018\,31\ldots - 0.073\,26\ldots - 0.146\,52\ldots$$
$$-0.195\,36\ldots - 0.195\,36\ldots$$

$$= 0.3711\ldots$$

$$= 0.371, \text{correct to 3 significant figures.}$$

As you would expect, $P(\text{Type II error})$ *decreases as* λ *gets lower.*

If the plants are distributed at random then the Poisson distribution should be a good model. If the plants propagate by seeds which are distributed by the wind then they are likely to be randomly distributed. If, however, they spread by underground roots then the plants might occur in groups.

Example 8.4.2

The average number of flaws per 100 metre length of yarn produced by a machine has been found to be 7. After the machine has been serviced, the number, X, of flaws in the first 300 metres of yarn produced by the machine is 27.

(a) Carry out a two-tail hypothesis test at the 5% level to test whether the average number of flaws produced by the machine has changed.
(b) For what values of X would the null hypothesis be rejected?
(c) Estimate the actual significance level of the test.
(d) Estimate $P(\text{Type II error})$ if the average has changed to 10.

(a) Assuming that the flaws are produced independently and at random, a suitable model for X would be the Poisson distribution.

$H_0: \lambda = 7$, $H_1: \lambda \neq 7$, where λ is the mean number of flaws per 100 m length.

Under H_0, $X \sim \text{Po}(3 \times 7) = \text{Po}(21)$.

Since $21 > 15$, $X \sim \text{Po}(21)$ can be approximated by $Y \sim \text{N}(21,21)$.

$$P(X \geqslant 27) = P(Y \geqslant 26.5) = P\left(Z \geqslant \frac{26.5 - 21}{\sqrt{21}}\right) = P(Z \geqslant 1.200\ldots).$$

The rejection region for a two-tail test at the 5% significance level is $|Z| \geqslant 1.96$. Since $1.200\ldots < 1.96$ there is insufficient evidence to say that the number of flaws has changed.

(b) The null hypothesis is rejected for $|Z| \geqslant 1.96$. The corresponding rejection region for X is given by

$$\frac{X - 0.5 - 21}{\sqrt{21}} \geqslant 1.96 \quad \text{and} \quad \frac{X + 0.5 - 21}{\sqrt{21}} \leqslant -1.96.$$

Solving these inequalities gives $X \geqslant 30.38\ldots$ and $X \leqslant 11.51\ldots$.

Since X must be an integer, the rejection region is taken as $X \geqslant 31$ and $X \leqslant 11$.

(c) The actual significance level of the test is equal to

$$P(X \geqslant 31 \text{ or } X \leqslant 11 \,|\, X \sim \text{Po}(21)).$$

Again using the normal approximation $Y \sim N(21,21)$,

$$P(X \geqslant 31) = P(Y \geqslant 30.5) = P\left(Z \geqslant \frac{30.5 - 21}{\sqrt{21}}\right) = P(Z \geqslant 2.073)$$
$$= 1 - \Phi(2.073)$$
$$= 1 - 0.9809 = 0.0191.$$

By symmetry, $P(X \leqslant 11) = 0.0191$.

Therefore the actual significance level of the test is $2 \times 0.0191 = 0.0382$.

(d) When $\lambda = 10$, $P(\text{Type II error}) = P(11 < X < 31 \,|\, X \sim \text{Po}(30))$.

Approximating $X \sim \text{Po}(30)$ by $Y \sim N(30,30)$,

$$P(11 < X < 31) = P(11.5 < Y < 30.5) = P\left(\frac{11.5 - 30}{\sqrt{30}} < Z < \frac{30.5 - 30}{\sqrt{30}}\right)$$
$$= P(-3.378 < Z < 0.091)$$
$$= \Phi(0.091) - (1 - \Phi(3.378))$$
$$= 0.5363 - (1 - 1.000)$$
$$= 0.5363.$$

The value of $\Phi(3.378)$ is outside the range of the tables on page 165; its value is taken to be 1.000.

Exercise 8C

1 The random variable X has a Poisson distribution with mean 6.5. A single observation of X is made and a test at a nominal significance level of 10% is carried out of whether the mean has decreased. If $X > 2$, it is assumed that the mean is unchanged. Find

(a) the probability of making a Type I error in the test,

(b) the probability of making a Type II error when the mean is actually 5.5.

Given that the observed value of X was 4, state the conclusion of the test and the type of error that was made.

2 The number of vehicle accidents at a cross-roads controlled by traffic lights may be modelled by a Poisson distribution with mean 1.5 per week. After a change in timing of the lights there were 6 accidents during the next 8 weeks. Is this evidence, at a 5% significance level, of a reduction in the weekly mean number of accidents?

What type of error might have been made in the test?

3 The proportion of all patients given laser surgery to treat astigmatism and short-sightedness and who suffer complications is reported to be, on average, 1 in 20. A newly formed company who give this treatment has given concern to a monitoring agency because the number of reported complications appears to be high. Records of the first 60 patients treated by the company are obtained and the agency will carry out a test, at a 10% significance level, of whether the true proportion of patients suffering complications is $\frac{1}{20}$. The number of patients who suffer complications is denoted by N.

(a) Explain why the distribution of N can be approximated by a Poisson distribution.

(b) State, with a reason, whether the agency should carry out a one-tail or a two-tail test.

(c) Show that the null hypothesis will be rejected if $N = 6$, but not if $N = 5$.

(d) State the type of error that might be made in the cases (i) $N = 4$, (ii) $N = 10$.

(e) Estimate the probability of making a Type II error in the test when the actual proportion of complications attributed to the company is $\frac{3}{40}$.

4 The number of times that a printing machine stops for attention during a given week has a Poisson distribution with mean 3.7. The machine undergoes some intensive adjustment and a two-tail test is carried out, based on the total number of stoppages, X, that occur over a period of 6 weeks. The test is of whether the mean number of stoppages per week has changed. The nominal significance level of the test is 5%.

(a) Find, in terms of X, the rejection region of the test.

(b) Estimate the actual significance level of the test.

(c) State the conclusion of the test for the case $X = 18$.

(d) Estimate the probability of making a Type II error when the actual mean number of weekly stoppages (after the adjustments) is 4.0.

5 The number of cars sold by staff at a car salesroom averages 0.8 per day. State what must be assumed for the number of sales made on a randomly chosen day to have a Poisson distribution.

Members of staff are given some extra training and the total sales over a period of 20 days after the training is denoted by T. It may be assumed that T has a Poisson distribution.

A test of whether the daily mean has increased is carried out at a significance level as close as possible to 5%.

(a) Find the rejection region of the test.

(b) Estimate the true significance level of the test.

(c) Estimate the probability of making a Type II error in the test when the mean number of daily sales is actually 0.9. Comment on its value.

6 The number of red cells in a small standard volume of blood of a healthy person is modelled by a Poisson distribution with mean 20. A doctor suspects that Rani has an abnormally high red cell count so she is given a blood test. The number of red cells in a standard volume of her blood is denoted by R. A statistical test of whether the doctor's suspicion is confirmed is carried out. If $R \le 25$, Rani's blood count is accepted as normal.

(a) Estimate the probability of making a Type I error in the test.

(b) Estimate the probability of making a Type II error when the mean is actually 30.

Miscellaneous exercise 8

1 In a significance test of a population mean μ, the null hypothesis $\mu = 0.3$ is tested against the alternative hypothesis $\mu \neq 0.3$ at the 10% significance level.

    (a) State, if possible, the probability of making a Type I error. If it is not possible then give a reason.

    (b) State, giving a reason, whether it is possible for both a Type I error and a Type II error to occur in the test.

    (c) State the type of error that might occur in the test given that the null hypothesis is rejected.

2 The number of misprints in mathematics books of about 300 pages published by a certain publisher (not this one, we hope!) has a distribution which can be modelled by a Poisson distribution with mean 4.8. A new director of publishing was appointed and in the first year of the director's appointment 5 mathematics books of about 300 pages were published. One of these was selected at random and a careful check found m misprints.

    (a) Find the rejection region of a test, at a nominal 5% significance level, of whether there is evidence of a reduction in the mean number of misprints per book.

    (b) For the case $m = 1$, state the conclusion of the test.

    (c) Given that the mean number of misprints had actually fallen from 4.8 to 0.5, find the probability of making a Type II error in the test.

3 A certain fly spray is known to kill at least 90% of flies on which it is used. A modification is made to the spray which a researcher believes will kill fewer than 90%. The actual proportion of flies killed by the modified spray is denoted by p. The spray is applied to 200 flies and kills k of them.

    (a) Use a suitable approximation (which should be justified) to find the set of values of k for which it would be accepted that $p < 0.9$ at a nominal $2\frac{1}{2}$% significance level.

    (b) Explain why the significance level cannot be exactly $2\frac{1}{2}$% and estimate the true significance level of the test.

    (c) The null hypothesis was accepted when the modified spray actually killed 85% of all flies on which it was used. State the type of error that occurred and calculate the probability of making that type of error.

4 The mass of silver contained in a particular brand of water purifying tablet is normally distributed with mean μ micrograms and standard deviation 0.52 micrograms. The null hypothesis $\mu = 100$ is tested against the alternative hypothesis $\mu \neq 100$ by taking a random sample of 10 tablets. The null hypothesis is accepted when $|\overline{X} - 100| < 0.27$, where $\overline{X}$ is the mean of the sample of 10 tablets.

    (a) Find the significance level of the test.

    (b) Calculate the probability of a Type II error for a sample of this size if μ changes from 100 to 99.5.

5 On a national railway system the number of signals passed at danger (SPADs) averages 19 per month. As a result of public anxiety, a vigorous safety campaign is mounted by the railway authorities with the aim of reducing this figure. A test is to be carried out, at the 5% significance level, of whether the average number of SPADs has decreased. The test statistic is X, the number of SPADs in the month following the campaign.

Using a suitable approximation, which you should state:

(a) find the rejection region for X;

(b) estimate the probability of a Type I error in this test.

In fact the average number of SPADs per month has decreased to 16.

(c) Estimate the probability of a Type II error in this test.

6 The lengths of a component used in the construction of a model aeroplane are being checked. Each of a random sample of 200 of the components, selected from a large batch, is measured and the lengths, x mm, are summarised by $\sum x = 1484.2$ and $\sum x^2 = 11\,098.19$.

(a) Calculate an unbiased estimate of the variance of the lengths of all components in the batch.

(b) State what is meant by 'an unbiased estimate'.

The components are produced in large batches and it is required that the mean length, μ mm, of the components in a batch should be at most 7.40 mm. Batches which do not meet this standard are rejected. The decision whether or not to reject a batch is based on a random sample of 200 components drawn from the batch. The mean length is denoted by $\overline{X}$.

(c) Find the rejection region of a test, at the 5% significance level, of whether or not a batch should be rejected.

(d) Find an approximation to the probability of making a Type II error when carrying out the test for the case when $\mu = 7.50$.

(e) Explain why (even without consideration of rounding errors) the answers to part (c) and part (d) are approximate.

7 For a statistics project Emma decides to investigate the possible existence of telepathy between her mother and herself. An experiment is set up using five differently coloured cards. One of these cards is chosen at random and shown to Emma's mother who concentrates on it for 20 seconds. At the same time, Emma concentrates and then names what she thinks is the colour of the chosen card. This is carried out 20 times altogether and on 7 occasions Emma names the correct colour.

(a) State suitable hypotheses, involving a probability, for a significance test that could indicate whether Emma is able to name the correct colour more often than would be expected by chance.

(b) Carry out the test at a nominal 5% significance level.

(c) State the type of error that might have been made in the conclusion of the test. Under what circumstances would this error not have been made?   (OCR, adapted)

8 Water used in a cooling tower is discharged into a river. Regulations require that the mean temperature of the discharged water should be no higher than 75 °C in order not to affect adversely the river's ecosystem. Samples of water are taken close to the cooling tower at random times over a period of a month and the temperatures recorded. The mean temperature of the water over this period is denoted by μ °C.

(a) State suitable hypotheses which can be used in a test of whether the regulations are observed.

(b) Describe, in the context of the question, the two types of error that might be made in the test.

(c) State, giving a reason, which type of error an environmentalist would consider to be less serious.

(d) For a particular sample of 30 readings the sample mean temperature was 76.16 °C with an estimated population variance of 2.62^2 °C^2. Carry out the test at the 10% significance level.

9 Factory-made dolls are assembled manually and the assembly times have a normal distribution with mean μ seconds and standard deviation 2.8 seconds. The usual method of assembly has $\mu = 12.4$ and in order to reduce this figure a supervisor has suggested a new method of assembly. A random sample of 10 workers will use the new method to assemble 5 dolls. If the sample mean time for assembling the 50 dolls is less than 11.4 seconds then the supervisor will decide to adopt the new method.

(a) State suitable hypotheses for the test implied by the supervisor's decision, and calculate the significance level of this test. It may be assumed that the standard deviation is 2.8 seconds.

(b) Calculate the probability of making a Type II error in the test when, in fact, $\mu = 10.0$.

(c) Find the smallest sample size that would have to be used so that, with the same decision procedure, the probability β of making a Type II error is smaller than 0.1. Find the value of β in this case.

10* A random variable X has a normal distribution with mean μ and variance 25. The null hypothesis $\mu = 20$ is tested against the alternative hypothesis $\mu < 20$ at the 10% significance level using the mean $\overline{X}$ of a random sample of n observations of X.

(a) Find, in terms of $\overline{X}$ and n, the rejection region of the test.

(b) Show that the probability β of making a Type II error in the test, when the actual value of μ is 19, is given by $\beta = 1 - \Phi\left(\frac{1}{5}\left(\sqrt{n} - 6.41\right)\right)$.

(c) Evaluate β when $n = 225$.

(d) What happens to the value of β as n increases?

(e) Find the value of n for which β is as close as possible to 0.01.

Revision exercise

1 A random variable X has probability density function $f(x)$ given by
$$f(x) = \begin{cases} kx(1-x) & \text{for } 0 \leqslant x \leqslant 1, \\ 0 & \text{otherwise.} \end{cases}$$

Show that the mean and variance of X are 0.5 and 0.05 respectively.

Find the probability that an observation chosen at random from this distribution is more than two standard deviations from the mean. (OCR)

2 From a large crop of potatoes, a random sample of 100 potatoes was washed and weighed. The weights, x grams, are summarised by $\sum x = 15\,829$ and $\sum x^2 = 2\,542\,324$.

(a) Calculate unbiased estimates of the mean and variance of the weight of all the potatoes in the crop.

(b) Calculate a 95% confidence interval for the mean weight of all the potatoes in the crop.

State, giving a reason, whether it is necessary to assume that the population is distributed normally in order for your calculation to be valid.

3 A multiple choice test consists of 50 questions each with 5 possible responses, only one of which is correct. A correct response receives a score of $+5$ and an incorrect response receives a score of $-a$. A particular candidate has not revised and does not know any of the answers for certain and so chooses a response at random for the 50 questions. The number of the candidate's correct responses is denoted by X and the total score for the paper by Y.

(a) State the distribution of X.

(b) Show that $Y = (5+a)X - 50a$.

(c) Find the value of a for which the expected score is 10 for a candidate who chooses all responses at random.

4 The lengths in millimetres of nine screws selected at random from a large consignment are found to be $7.99, 8.01, 8.00, 8.02, 8.03, 7.99, 8.00, 8.01, 8.01$.

(a) Calculate unbiased estimates of the population mean and variance.

(b) Assuming a normal distribution with variance 0.0001, test, at the 5% level, the hypothesis that the population mean is 8.00 against the alternative hypothesis that the population mean is not 8.00. (OCR)

5 Along a particular stretch of road, vehicle breakdowns occur over time at an average rate of 3.2 per day in the northbound direction and 2.8 per day along the southbound direction. Estimate the probability that a total of between 20 and 25 (inclusive) breakdowns occur along this stretch of road over a randomly chosen period of three days.

6 What do you understand by the phrase 'an unbiased estimate of population variance'?

(a) An infinite population consists of the numbers $1, 2, 3$ in equal proportion. Write down the population mean, and hence calculate the population variance.

(b) Random samples of size 2 are obtained from the population. Write down all possible samples of size 2. Hence write down the probability distribution of the means of samples of size 2. Obtain the expected value of the mean and determine if the 'sample mean' is unbiased in this case.

(c) Obtain the distribution of sample variance and find the expected value of sample variance. Demonstrate that the sample variance is not an unbiased estimator of population variance in this case. Show also that

$$\frac{n}{n-1} \times (\text{sample variance})$$

is an unbiased estimator in this case, where n is the sample size.   (OCR)

7 Statistical investigations may involve random sampling. Explain what is meant by the term 'random sampling' and state why random sampling is used.

A student, when asked how to choose a 10% sample from the 150 houses in a village, gave the following reply:

'Number the houses from 1 to 150. Choose 15 two-digit random numbers. This gives the 10% sample.'

Criticise this student's method. Say how to improve it. Demonstrate your improved version by using the following random numbers to select three houses from the 150 in the village.

23895   40522   02568   (OCR)

8 The mass of luggage that aircraft passengers take with them is normally distributed with mean 20 kg and standard deviation 5 kg. A certain type of aircraft carries 100 passengers.

(a) Find the probability that the mean mass of the passengers' luggage exceeds 21.3 kg.

(b) To what extent does your answer depend on the distribution of the individual masses being normal?   (OCR)

9 A continuous random variable X has a probability density function given by

$$f(x) = \begin{cases} a - \frac{1}{2}(x-1)^2 & \text{for } 0 \le x \le 2, \\ 0 & \text{otherwise.} \end{cases}$$

Calculate

(a) the value of a,

(b) μ, the mean of the distribution,

(c) σ^2, the variance of the distribution.   (OCR, adapted)

10 In a study of the use by university students of the internet, a random sample of 500 students was selected and 184 were found to spend longer than 3 hours each week using the internet. Calculate a symmetric 95% confidence interval for the proportion of all university students who spend longer than 3 hours each week using the internet.

11 A machine is designed to produce rods 2 cm long with standard deviation 0.02 cm. The lengths may be taken as normally distributed. The machine is moved to a new position in the factory and, in order to check whether the setting for the mean length has altered, the lengths of a sample of 9 rods are measured. The standard deviation may be considered to be unchanged. The lengths of the 9 rods in the sample are given below.

   2.04   1.97   1.99   2.03   2.04   2.10   2.01   1.98   2.07

(a) Test, at the 5% level, whether the machine's setting has been altered. State your hypotheses and conclusions clearly.

(b) If, in fact, the setting of the machine has been altered so that the mean length of rods produced is now 2.01, find the probability of making a Type II error.   (OCR)

12 Potholes in roadways occur at random. Give two reasons why a Poisson distribution may be an appropriate model for the number of potholes in a randomly selected 100 m stretch of roadway.

Records held by the highways department of county C indicate that the mean number of potholes in such stretches of roadway is 2.5.

(a) Calculate the probability of there being between 3 and 5 (inclusive) potholes in a randomly chosen 100 m stretch of roadway in county C.

In a neighbouring county, D, there are on average 1.8 potholes per 100 m of roadway.

(b) Calculate the probability that there are fewer than 3 potholes in each of 5 randomly chosen 200 m stretches of county D's roadways.

(c) Use a normal approximation to find the probability that in a randomly chosen 1000 m stretch of county D's roadways there are fewer than 15 potholes.   (OCR, adapted)

13 The continuous random variable X has probability density function given by

$$f(x) = \begin{cases} k(1+x^2) & \text{for } -1 \leqslant x \leqslant 1, \\ 0 & \text{otherwise,} \end{cases}$$

where k is a constant.

(a) Find the value of k.

(b) Determine

   (i) $E(X)$,

   (ii) $Var(X)$.   (OCR)

14 On the surface of postage stamps there are either one or two phosphor bands. 90% of stamps have two bands and the rest have one band. Of those having one band, 95% have the band in the centre of the stamp and the remainder have the band on the left edge of the stamp.

(a) Determine, using a normal approximation, the probability that in a random sample of 100 stamps there are between 5 and 15 (inclusive) having one phosphor band.

(b) Determine, using a Poisson distribution, the probability that in a random sample of 100 stamps there are fewer than 3 stamps which have only a single band, this band being on the left edge of the stamp.   (OCR)

15 From past experience, it is known that the time a customer took to be served in a Post Office was a normal variable with mean 6.2 minutes and standard deviation 2.1 minutes.

A new queuing system was introduced by the manager and after this a survey of 20 customers was carried out to see whether there was a decrease in the time it took for a customer to be served. Let T be the time in minutes for a randomly chosen customer to be served after the new queuing system has been introduced and let t_1, t_2, ..., t_{20} be the random sample of 20 observations of T. The results may be summarised as

$$\sum t = 108.4.$$

The standard deviation of T is assumed to be 2.1 minutes.

(a) Test, at the 5% level, the hypothesis that the mean time for a customer to be served has reduced after the introduction of the new queuing system.

(b) For the test given in part (a) calculate the probability of making a Type II error in the case when the true mean time to be served is 5.8 minutes. (OCR)

16 The playing times of a particular brand of audio tape are normally distributed with mean μ minutes and standard deviation 0.24 minutes. The manufacturer states that $\mu = 60$. A large batch of these tapes is delivered to a store and, in order to check the manufacturer's statement, the playing times of a random sample of 10 tapes is tested. The null hypothesis $\mu = 60$ is tested against the alternative hypothesis $\mu < 60$ at the 1% significance level.

(a) Find the range of values of the sample mean $\overline{X}$ for which the null hypothesis is rejected, giving 2 decimal places in your answer.

(b) State what 'a Type II error has occurred' means in the context of the playing times of tapes.

(c) Calculate the probability of making a Type II error when, in fact, $\mu = 59.7$.

17 In a random sample of 250 families who watched TV last Thursday, the percentage who watched a particular game show is denoted by p_s. For all such random samples, state, giving a reason, whether the random variable P_s is discrete or continuous.

For a particular sample $p_s = 34\%$.

(a) Calculate a symmetric 95% confidence interval for p, the percentage of all viewing families who watched the game show last Thursday.

(b) Give a reason, apart from rounding, why your interval is approximate.

A further random sample of families who watched TV last Thursday was selected. Estimate the sample size for which a symmetric 99% confidence interval for p_s would have a width of 5%.

What can be said about the accuracy of p_s, found using the calculated sample size, as an estimate of p?

18 A servicing engineer finds that the number of jobs he completes in a working session has a Poisson distribution with mean 4. If the sessions are independent, what is the distribution of the number of jobs he completes in n sessions, and how may this be approximated when n is large?

If he has 100 jobs to do, how many sessions should he allow in order to be 95% sure that he will be able to complete them all? (OCR)

19 The process of manufacturing a certain kind of dinner plate results in a proportion 0.13 of faulty plates. An alteration is made to the process which is intended to reduce the proportion of faulty plates. State suitable null and alternative hypotheses for a statistical test of the effectiveness of the alteration.

In order to carry out the test, the quality control department counts the number of faulty plates in a random sample of 2500. If 290 or fewer faulty plates are found then it will be accepted that the alteration does result in a reduction in the proportion of faulty plates. Calculate the significance level of this test, using a suitable normal approximation.

Calculate the probability of making a Type II error in the above test, given that the alteration results in a decrease in the proportion of faulty plates to 0.11.

20 A manufacturer of a new soap powder wishes to predict the likely volume of sales in a town. Four different methods are suggested for selecting people for a questionnaire. Discuss the advantages and disadvantages of each method and select the one which you think is the best, giving reasons for your choice.

(a) Take every 20th name on the electoral register of the town.

(b) Choose people leaving a supermarket in the town, ensuring that the numbers in each gender, age and social class category are proportional to the numbers in the population.

(c) Select houses at random from a town plan and interview one person from each house.

(d) Choose at random one name from each page of the telephone directory and ring them up. (OCR, adapted)

21 A random variable X has a $N(15,9)$ distribution. A random sample of 5 observations of this distribution is to be taken. The mean of the 5 observations is denoted by $\overline{X}$.

(a) State the distribution of $\overline{X}$ and give its mean and variance.

(b) Calculate the probability that $\overline{X}$ is less than 17.

A random variable Y has mean 7 and variance 20. A random sample of 100 observations of this distribution is to be taken. The mean of the 100 observations is denoted by $\overline{Y}$.

(c) Give the mean and variance of $\overline{Y}$.

(d) State the approximate distribution of $\overline{Y}$ and give the name of the theorem which you have used.

(e) Calculate the probability $P(6.9 < \overline{Y} < 7.1)$.

22 When cars arrive at a certain T-junction they turn either right or left. Part of a study of road usage involved deciding between the following alternatives.

Cars are equally likely to turn right or left.

Cars are more likely to turn right than left.

(a) State suitable null and alternative hypotheses, involving a probability, for a significance test.

(b) Out of a random sample of 40 cars, n turned right. Use a suitable approximation to find the least value of n for which the null hypothesis will be rejected at the 2% significance level.

(c) For the test described in part (b), calculate the probability of making a Type II error when, in fact, 80% of all cars arriving at the junction turn right.

23 A researcher is studying a museum's collection of tetradrachms coined during the reign of Alexander III ('the Great'; 336–323 B.C.). Each of a random sample of 50 coins was weighed and their masses, x grams, are summarised by

$$\sum x = 848.27, \qquad \sum x^2 = 14\ 396.1015.$$

(a) Use an unbiased estimate of variance to calculate an approximate 99% confidence interval for the mean mass (in grams) of all tetradrachms coined during Alexander's reign, giving the end-values of the interval to 2 decimal places.

(b) Estimate the size of the random sample of coins that would be required to give a 95% confidence interval whose width is half that of the interval calculated in part (a).

(c) It was found later that the scales were consistently underweighing by 0.05 grams. State which of the results of part (a) and (b) should be amended and which should not. Give the amended values.

24 The total number of units of electricity that Kallie uses over the period April to June in any year has a normal distribution with mean 920 and standard deviation 95. The supplier has two methods of payment. In method 1, there is a fixed charge of $20 together with a charge of 12.90 cents per unit of electricity used. Method 2 has no fixed charge but each unit of electricity used costs 15.76 cents. Calculate the probability that, over the same period next year, Kallie would pay more using method 1 than using method 2.

Over the same period, the units of electricity that Leroy uses have a normal distribution with mean 750 and standard deviation 64. Given that Leroy uses method 2 and Kallie uses method 1, calculate the probability that, over the period April to June next year, Leroy's electricity bill will be greater than Kallie's. Assume that the charges are unchanged, that the amounts of electricity used follow the same distributions as before and that the amounts of electricity used by Kallie and Leroy are independent.

Practice examination 1

Time 1 hour 15 minutes

Answer all the questions.
The use of an electronic calculator is expected, where appropriate.

1 The random variable X has the distribution $N(40,25)$, and $\overline{X}$ denotes the mean of a random sample of 10 observations of X. Find $P(\overline{X} < 42)$. [4]

2 Statistical investigations may involve random sampling.

   (i)   Explain what is meant by the term 'random sampling', and state why random sampling is used. [3]

   (ii) A random sample of 5 students is to be taken from the 763 students who attend a college. Explain briefly how this could be done by using random numbers. [3]

3 A student is investigating the shape of a certain type of shell found on the seashore. She makes several measurements for each shell and combines the results into a single 'shape index', x. The resulting values from a random sample of 208 shells are summarised by $\Sigma x = 633.36$, $\Sigma x^2 = 2640.4612$.

   (i)   Calculate unbiased estimates of the population mean and variance for the shape index of shells of this type. [3]

   (ii) Obtain a symmetric 90% confidence interval for the population mean. [4]

4 The continuous random variable X has probability density function given by

$$f(x) = \begin{cases} k(3-x)^2 & \text{for } 0 \leqslant x \leqslant 3, \\ 0 & \text{otherwise,} \end{cases}$$

   where k is a constant.

   (i)   Show that $k = \frac{1}{9}$. [2]

   (ii) Find $P(1 \leqslant X \leqslant 2)$. [2]

   (iii) Find $E(X)$. [3]

5 It is thought that more baby boys than baby girls are being born. A test of the null hypothesis $p = 0.5$ against the alternative hypothesis $p > 0.5$ is carried out, where p denotes the probability of a randomly chosen baby being male. For the test, a random sample of 10 babies is taken and the null hypothesis is rejected if 8, 9 or 10 of them are male.

   (i)   Calculate the probability of a Type I error in this test. [3]

   (ii) Calculate the probability of a Type II error if the true value of p is 0.6. [4]

6 The accident and emergency department at a city hospital keeps records of the numbers of cases arriving between the hours of 2200 and 2300 throughout the week. The numbers, each day, have independent Poisson distributions with mean 4.3 for Monday to Friday and mean 6.2 for Saturday and Sunday.

(i)   Calculate the probability that, on a particular Wednesday, at least 4 cases will arrive between 2200 and 2300 hours.   [3]

The total number of cases that arrive between 2200 and 2300 hours during one randomly chosen week is denoted by T.

(ii) State the probability distribution of T.   [2]

(iii) Use a suitable approximation to find $P(30 \leqslant T \leqslant 40)$.   [4]

7 In a certain population, the weights in kg of men and women have independent normal distributions with means and standard deviations as follows.

> Men: mean 75,   standard deviation 6.4.
> Women: mean 54,   standard deviation 4.9.

One man and one woman are chosen at random.

(i)   Find the probability that their total weight is greater than 140 kg.   [4]

(ii) Find the probability that the woman's weight is less than half the man's weight.   [6]

Practice examination 2

Time 1 hour 15 minutes

Answer all the questions.
The use of an electronic calculator is expected, where appropriate.

1 In a random sample of 200 people who listened to the radio on one particular evening, the percentage who listened to a certain current affairs program was 28%. Calculate an approximate 99% confidence interval for the population proportion of listeners who listened to this program. [4]

2 A six-sided dice is suspected of being less likely to show a 'six' than any of the other numbers. In an experiment to investigate this suspicion, the dice is thrown 20 times, and only one 'six' is obtained. State appropriate null and alternative hypotheses, and carry out a hypothesis test at the 10% significance level to test the suspicion. [5]

3 The continuous random variable X has a normal distribution with mean 20 and standard deviation 3. The mean of a random sample of n observations of X is denoted by $\overline{X}$. Given that $P(\overline{X} > 21)$ is between 0.01 and 0.05, find the set of possible values of n. [6]

4 A shopkeeper knows that the mean number of CD players sold per day in his shop is 3.
   (i) State a probability distribution which you might expect to be a reasonable model for the number of CD players sold by the shopkeeper on a randomly chosen day. [1]
   (ii) Hence calculate the probability that the shopkeeper sells exactly 5 CD players on a randomly chosen day. [2]
   (iii) The shopkeeper opens his shop for 6 days every week. Use a suitable approximation to determine the number of CD players which the shopkeeper should have in stock at the beginning of a randomly chosen week to be at least 95% certain of being able to meet the demand during the week. [4]

5 The amount of petrol, in tens of thousands of litres, sold in a week at a petrol filling station is modelled by the continuous random variable X with probability density function given by

$$f(x) = \begin{cases} kx(4 - x^2) & \text{for } 0 \leqslant x \leqslant 2, \\ 0 & \text{otherwise.} \end{cases}$$

   (i) Find the value of the constant k. [2]
   (ii) Calculate the probability that in a randomly chosen week the amount of petrol sold is more than 14 000 litres. [3]
   (iii) Show that the median, m, of X satisfies the equation
   $$m^4 - 8m^2 + 8 = 0,$$
   and hence find the median amount of petrol sold in a week. [3]

6 Robert and Sipho are athletes who specialise in the long jump. The lengths of their jumps may be modelled by normal distributions with means and variances as follows.

	Mean (m)	Variance (m^2)
Robert	8.13	0.24
Sipho	8.49	0.21

They both compete in an athletics competition, in which they each have three jumps. The length of any jump may be assumed to be independent of all other jumps. Calculate the probability that

(i)   Robert and Sipho's first jumps differ by more than 1 m,   [5]

(ii)   the sum of Robert's three jumps exceeds the sum of Sipho's three jumps.   [4]

7 A machine produces metal rods, whose lengths are normally distributed with standard deviation 0.1 cm. The machine is set up for the rods to have a mean length of 2 cm. To check whether the setting is accurate, a random sample of 20 rods is taken, and the lengths, x cm, are measured. It is found that $\Sigma(x-2) = 0.84$.

(i)   Test, at the 5% significance level, whether the machine has been set up correctly. State your hypotheses and conclusions clearly.   [5]

(ii)   Find the probability of making a Type II error in this test if the mean length of rods produced by the machine is in fact 2.05 cm.   [6]

The Normal Distribution Function

If Z has a normal distribution with mean 0 and variance 1 then, for each value of z the table gives the value of $\Phi(z)$, where

$$\Phi(z) = P(Z \leqslant z).$$

For negative values of z use $\Phi(-z) = 1 - \Phi(z)$.

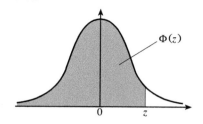

z	0	1	2	3	4	5	6	7	8	9	1	2	3	4	5	6	7	8	9
														ADD					
0.0	0.5000	0.5040	0.5080	0.5120	0.5160	0.5199	0.5239	0.5279	0.5319	0.5359	4	8	12	16	20	24	28	32	36
0.1	0.5398	0.5438	0.5478	0.5517	0.5557	0.5596	0.5636	0.5675	0.5714	0.5753	4	8	12	16	20	24	28	32	36
0.2	0.5793	0.5832	0.5871	0.5910	0.5948	0.5987	0.6026	0.6064	0.6103	0.6141	4	8	12	15	19	23	27	31	35
0.3	0.6179	0.6217	0.6255	0.6293	0.6331	0.6368	0.6406	0.6443	0.6480	0.6517	4	7	11	14	18	22	25	29	32
0.4	0.6554	0.6591	0.6628	0.6664	0.6700	0.6736	0.6772	0.6808	0.6844	0.6879	4	7	11	14	18	22	25	29	32
0.5	0.6915	0.6950	0.6985	0.7019	0.7054	0.7088	0.7123	0.7157	0.7190	0.7224	3	7	10	14	17	20	24	27	31
0.6	0.7257	0.7291	0.7324	0.7357	0.7389	0.7422	0.7454	0.7486	0.7517	0.7549	3	7	10	13	16	19	23	26	29
0.7	0.7580	0.7611	0.7642	0.7673	0.7704	0.7734	0.7764	0.7794	0.7823	0.7852	3	6	9	12	15	18	21	24	27
0.8	0.7881	0.7910	0.7939	0.7967	0.7995	0.8023	0.8051	0.8078	0.8106	0.8133	3	5	8	11	14	16	19	22	25
0.9	0.8159	0.8186	0.8212	0.8238	0.8264	0.8289	0.8315	0.8340	0.8365	0.8389	3	5	8	10	13	15	18	20	23
1.0	0.8413	0.8438	0.8461	0.8485	0.8508	0.8531	0.8554	0.8577	0.8599	0.8621	2	5	7	9	12	14	16	19	21
1.1	0.8643	0.8665	0.8686	0.8708	0.8729	0.8749	0.8770	0.8790	0.8810	0.8830	2	4	6	8	10	12	14	16	18
1.2	0.8849	0.8869	0.8888	0.8907	0.8925	0.8944	0.8962	0.8980	0.8997	0.9015	2	4	6	7	9	11	13	15	17
1.3	0.9032	0.9049	0.9066	0.9082	0.9099	0.9115	0.9131	0.9147	0.9162	0.9177	2	3	5	6	8	10	11	13	14
1.4	0.9192	0.9207	0.9222	0.9236	0.9251	0.9265	0.9279	0.9292	0.9306	0.9319	1	3	4	6	7	8	10	11	13
1.5	0.9332	0.9345	0.9357	0.9370	0.9382	0.9394	0.9406	0.9418	0.9429	0.9441	1	2	4	5	6	7	8	10	11
1.6	0.9452	0.9463	0.9474	0.9484	0.9495	0.9505	0.9515	0.9525	0.9535	0.9545	1	2	3	4	5	6	7	8	9
1.7	0.9554	0.9564	0.9573	0.9582	0.9591	0.9599	0.9608	0.9616	0.9625	0.9633	1	2	3	4	4	5	6	7	8
1.8	0.9641	0.9649	0.9656	0.9664	0.9671	0.9678	0.9686	0.9693	0.9699	0.9706	1	1	2	3	4	4	5	6	6
1.9	0.9713	0.9719	0.9726	0.9732	0.9738	0.9744	0.9750	0.9756	0.9761	0.9767	1	1	2	2	3	4	4	5	5
2.0	0.9772	0.9778	0.9783	0.9788	0.9793	0.9798	0.9803	0.9808	0.9812	0.9817	0	1	1	2	2	3	3	4	4
2.1	0.9821	0.9826	0.9830	0.9834	0.9838	0.9842	0.9846	0.9850	0.9854	0.9857	0	1	1	2	2	2	3	3	4
2.2	0.9861	0.9864	0.9868	0.9871	0.9875	0.9878	0.9881	0.9884	0.9887	0.9890	0	1	1	1	2	2	2	3	3
2.3	0.9893	0.9896	0.9898	0.9901	0.9904	0.9906	0.9909	0.9911	0.9913	0.9916	0	1	1	1	1	2	2	2	2
2.4	0.9918	0.9920	0.9922	0.9925	0.9927	0.9929	0.9931	0.9932	0.9934	0.9936	0	0	1	1	1	1	1	2	2
2.5	0.9938	0.9940	0.9941	0.9943	0.9945	0.9946	0.9948	0.9949	0.9951	0.9952	0	0	0	1	1	1	1	1	1
2.6	0.9953	0.9955	0.9956	0.9957	0.9959	0.9960	0.9961	0.9962	0.9963	0.9964	0	0	0	0	1	1	1	1	1
2.7	0.9965	0.9966	0.9967	0.9968	0.9969	0.9970	0.9971	0.9972	0.9973	0.9974	0	0	0	0	0	1	1	1	1
2.8	0.9974	0.9975	0.9976	0.9977	0.9977	0.9978	0.9979	0.9979	0.9980	0.9981	0	0	0	0	0	0	0	1	1
2.9	0.9981	0.9982	0.9982	0.9983	0.9984	0.9984	0.9985	0.9985	0.9986	0.9986	0	0	0	0	0	0	0	0	0

Critical values for the normal distribution

If Z has a normal distribution with mean 0 and variance 1 then, for each value of p, the table gives the value of z such that $P(Z \leqslant z) = p$.

p	0.75	0.90	0.95	0.975	0.99	0.995	0.9975	0.999	0.9995
z	0.674	1.282	1.645	1.960	2.326	2.576	2.807	3.090	3.291

Answers

Most non-exact numerical answers are given correct to 3 significant figures.

1 The Poisson distribution

Exercise 1A (page 3)

1 (a) 0.2240 (b) 0.1991 (c) 0.5768

2 (a) 0.2218 (b) 0.3696 (c) 0.8352

3 (a) 0.7787 (b) 0.6916 (c) 0.2090

4 (a) 0.1353 (b) 0.2707 (c) 0.2381

5 (a) 0.0821 (b) 0.5162

6 (a) 0.6065 (b) 0.4634

7 (a) 0.006 74 (b) 49.9 s

8 (a) 0.0916 (b) 0.206 particle s^{-1}

Exercise 1B (page 9)

1 (a) Yes (b) No (c) Yes
(d) Yes, provided claims are not caused by, say, freak weather conditions.

2 0.0005, 0.006 730; 0.000 005, 0.006 739; 0.000 000 05, 0.006 738
$e^{-5} = 0.006\ 738$. (The final answer in the table and e^{-5} agree to 8 decimal places.)

3 (a) 0.29, 0.39, 0.16, 0.13, 0.03
(b) 1.23, 1.21; Poisson is a suitable model.
(c) 0.29, 0.36, 0.22, 0.09, 0.03
(d) Supports comment in part (b).

4 1.71, 1.97;
0.12, 0.45, 0.21, 0.12, 0.05, 0.02, 0, 0.02
0.18, 0.31, 0.26, 0.15, 0.06, 0.02, 0.01, 0.002
Poisson is not suitable.

5 $\mu = 0.79$, $\sigma^2 = 0.87$
Theoretical probabilities 0.45, 0.36, 0.14, 0.04, 0.01, 0.001. Yes.

Exercise 1C (page 14)

1 (a) (i) 0.222 (ii) 0.221
(b) (i) 0.0966 (ii) 0.0965

2 (a) 0.8795 (b) 0.0077

3 (a) 0.108 (b) 0.0273

4 0.0190

5 (a) 0.953 (b) 0.434

6 (a) 0.904 (b) 0.0119 (c) 0.320

Exercise 1D (page 16)

1 (a) 0.608 (b) 0.178 (c) 0.212

2 (a) 0.265 (b) 0.0497

3 0.209

4 0.0442

5 (a) 0.689 (b) 0.0446

6 (a) 0.759 (b) 0.0599

7 62

8 123

Miscellaneous exercise 1 (page 17)

1 0.191

2 Because n is large (> 50) and p is small ($np < 5$); 0.1953

3 0.790; the calls occur randomly, independently, singly and at a constant rate

4 (a) 0.268 (b) 0.191

5 2; 0.188; 0.929

6 Y because $np = 4.2 < 5$; 0.210

7 The injuries occur randomly, independently, singly and at a constant rate.
0.5, 0.481; possibly Poisson since the mean and variance are approximately equal.
32, 16, 4, 1, 0; note that the frequencies do not add up to 52 because of rounding errors.

8 (a) 0.067 (b) 0.286 (c) 0.739
(d) 0.465
(e) Po(29.7); 0.442, using N(29.7, 29.7)

9 $\frac{25}{9}$; 0.240; 0.303

10 (a) (i) 0.741 (ii) 0.037 (c) 100

11 3; 18.5%

12 0.647; 0.185

2 Linear combinations of random variables

Exercise 2A (page 26)

1 (a) 1.3, 0.61 (b) 5.9, 5.49

2 (a) -4, 5 (b) 89, 80

3 1, 0.5

4 0.5, 0.25; 0.48, 0.16; 0.3, 0.37, 0.2, 0.04; 1, $\frac{1}{2}$, 0.8, 0.48, 1.8, 0.98

5 3.1, 3.69; 0.2, 0.16, 0.1, 0.14, 0.2

6 24, 12, 6, 24

7 8 mm, 0.14 mm^2

8 3.5, $\frac{35}{12}$, 3, 9, 6.5, $\frac{143}{12}$

9 $\frac{51}{8}$, $\frac{213}{64}$

10 82.5, 311

Exercise 2B (page 32)

1 0.0592

2 0.0638

3 0.0228

4 0.127

5 0.362

6 0.0317

7 $\mu = 16$, $\sigma^2 = 9$

8 (a) 5, 5   (b) 1, 5   (c) 11, 27
(a) is Poisson

9 0.116

10 0.242

Miscellaneous exercise 2 (page 33)

1 10, 50

2 2, 1; 4, 3

3 $\frac{1}{15}$, $\frac{2}{3}$, $\frac{34}{45}$; $\frac{4}{3}$, $\frac{68}{45}$

4 12 kg, 57 g, 3.97%, 765 g

5 (a) 0.189   (b) 0.308   (c) 0.184

6 0.121

7 (a) 0.132   (b) 0.228

8 (a) 0.655   (b) 0.314

9 (a) 0.681   (b) 0.113   (c) 0.159

10 (a) 0.309   (b) 0.235; 0 because
$$P(N > 3.5n) = P\left(z > \tfrac{1}{4}\sqrt{n}\right) = 1 - \Phi\left(\tfrac{1}{4}\sqrt{n}\right).$$

3 Continuous random variables

Exercise 3A (page 42)

1 (a) $\frac{1}{4}$ (b) $\frac{1}{16}$ (c) $\frac{3}{16}$

2 (a) $\frac{1}{9}$ (b) $\frac{8}{27}$ (c) 0.454   (d) 1.75

3 (a) $\frac{1}{15}$ (b) $\frac{11}{40}$

4 (a) $\frac{3}{32}$ (b) $\frac{1}{2}$ (c) $\frac{5}{32}$ (d) $\frac{5}{16}$
(e) $\frac{47}{128}$

5 (a) 2   (b) 0.135   (c) 0.632

6 (a) 400   (b) $\frac{4}{5}$ (c) $\frac{1}{3}$ (d) $\frac{1}{9}$

7 (a) 3, $\frac{1}{9}$ (b) $\frac{1}{8}$

Exercise 3B (page 45)

1 2.381

2 (a) 800 hours   (b) 1000

3 (b) $2 + \frac{1}{2}\sqrt{2}$ (c) $\frac{1}{2}\left(1 + \sqrt{3}\right)$

4 (b) 347 hours   (c) 52.7

5 (b) $5 - \frac{5}{2}\sqrt{2}$

Exercise 3C (page 49)

1 $\frac{9}{4}$, $\frac{27}{80}$

2 $\frac{8}{3}$, $\frac{1}{18}$

3 (b) $\frac{8}{3}$, $\frac{32}{9}$

4 (b) 2 kg, $\frac{4}{5}$ kg^2

5 (b) 15 hours, 75 hours^2

6 (c) 584 hours, 19 100 hours^2

7 500 hours, 250 hours^2

8 (b) $\dfrac{1}{k}$, $\dfrac{1}{k^2}$

Miscellaneous exercise 3 (page 50)

1 (a) $\frac{1}{4}$ (b) $\frac{8}{3}$

2 (a) 2.5   (b) 2.08

3 (a) $\frac{1}{50}$ (b) $\frac{10}{3}$ min   (c) $\frac{50}{9}$ min^2

4 (b) $\frac{1}{8}$ (c) 0.141   (d) $\frac{4}{3}$

5 (a) $\sqrt{2}$ (b) $\frac{4}{3}$ (c) $\frac{2}{9}$

6 (a) $f(u) = \frac{1}{2}$, for $0.5 \leqslant u \leqslant 2.5$, and 0 otherwise
(b) $\frac{3}{2}$ (c) $\frac{1}{3}$

7 (a) 0.6   (b) 0.5 m   (c) 0.03 m^2

8 (a) $\frac{1}{4}$ (b) 1.6   (c) 0.107   (d) 1.68
(e) 0.697

9 (a) $\frac{1}{64}$ (b) $\frac{16}{5}$ (c) 0.427   (d) $\frac{15}{256}$

10 (a) $\dfrac{1}{\ln 2}$ (b) $\dfrac{1}{\ln 2}$ (c) 0.083   (d) $\sqrt{2}$

11 (b) 0.330   (c) 0.368

12 (a) $\dfrac{\ln 2}{a}$ (b) 6 (to the nearest integer)

13 (b) $\frac{3}{2}a$

14 (b) 3 years   (c) 0.315
(d) On average the guarantee is cheaper.
(e) 0.5

4 Sampling

Exercise 4A (page 59)

1 (a) Biased; people are likely to approve.
(b) Each age group is represented, but not necessarily in proportion to age distribution.
(c) Satisfactory if chosen at a suitable time.

2 Each selection is independent and has an equal probability of being chosen.
  (a) Biased; not all people have a telephone.
  (b) Biased; only one area at one time is chosen.
  (c) Should be satisfactory.

3 (a) Biased; does not include people who buy the paper, say, from a street vendor.
  (b) Biased; respondents are only those who care enough to complete the form.

4 The number of words starting with the same initial letter is different for each letter.
For example, number the letters from 1 to 26 and use some form of random number generator.

Exercise 4B (page 71)

1 (a) False. $\overline{X}(n)$ has an *approximate* normal distribution for *large values* of n if X is not normally distributed.
  (b) True   (c) False. See part (a).
  (d) False. True for any n.

2 (a) N(2.4, 0.0432) (b) 0.3152
Neither of them depend on the central limit theorem.

3 $P(X = x) = \frac{1}{6}$, $x = 1, 2, 3, 4, 5, 6$
  (a) 0.155   (b) 0.845

4 (a) 0.271
  (b) 0.0021; the assumption is that the biology students at the seminar are a random sample of all fully grown males.
The second, because n is larger; the larger n, the closer the distribution of the mean is to a normal distribution.

5 (a) 0.968   (b) 95.54%

6 (a) 0.0084 (b) 82

7 0.9957

8 0.0434

9 (a) 0.981   (b) $1 - \Phi(-3.338) \approx 1$

10 $0.9615 > 95\%$

11 24.9; no, since X has a normal distribution.

Miscellaneous exercise 4 (page 73)

1 30

2 (a) 16 (b) 0.6554

3 (a) Sample restricted by length of throw. Satisfactory if researcher moves to different parts of the field before throwing.
  (b) Sample points restricted to centres of grid squares, also to points greater than 0.5 m from the boundary. However, sample should be satisfactory.

502

4 (a) $\frac{9}{7}$, $\frac{3}{49}$
  (b) 0.0299; distribution of T is very skewed so answer not very accurate.

5 (a) 0.0594 (b) 0.0023

6 (a) $2.7, \dfrac{2.7}{n}$ (b) 0.0192

7 (a) $N\left(50, \dfrac{64}{n}\right)$ (b) 0.0062
  (c) 22
Approximately normal if n is large, but nothing if n is small.

8 (a) $p = 0.8$, $n = 80$ (b) 0.037

9 2.11 hours

10 0.010

11 (8.49, 9.51);
(7.75, 10.25); (Hint: $k = 5$); the interval is wider, probability greater than 0.96.

5 Estimation

Exercise 5A (page 84)

1 $1.83, 0.119 \$^2$

2 (a) $0.013\,775\,\text{cm}^2$ (b) $0.0145\,\text{cm}^2$

3 (a) 2.49, 1.61
  (b) 0.182; month may give a biased sample, for example, in the holiday season traffic may be heavier and accidents more likely.

4 (a) 50.633 g, 147.365 g^2 (b) 0.37

5 5.6, 2.569, 0.0248; variance approximate and normal approximation used.

6 101.1 g, 1.369 g^2, $\Phi(2.973) \approx 1$; it assumes that sample parameters are good approximations to population parameters.

Exercise 5B (page 91)

1 $[496.5, 504.5]$, 190

2 (a) $[0.981, 1.028]$ (b) 0.0233 litres

3 (a) 5.155   (b) 8.48   (c) 10.10

4 $[169.4, 181.0]$; no, 178 is within the interval.
It is assumed that the biology students at the seminar are a random sample of all fully-grown male biology students.

5 (a) $[5.013, 5.057]$
  (b) 5.00 cm is outside the interval, indicating that the mean has increased.

6 (a) $[15.67, 16.17]$ (b) Six more

Exercise 5C (page 94)

1 (a) $[4.73, 5.62]$

(b) Assume sample mean has a normal distribution, which is justified by the central limit theorem (sample size large); also that the sample variance is a good estimate of the population variance which is justified by the large sample size.

2 (a) 498.1 g, 4.839 g^2 (b) $[497.7, 498.5]$

The confidence interval will not contain the mean 5% of the time.

3 (a) 5.968 m^2 (b) $[6.44, 8.22]$
$[5.10, 6.88]$

4 (a) $2.43, 2.288$ (b) $[2.18, 2.68]$

5 (a) $[84.1, 89.1]$ (b) Interval is only valid for mean of 8-year-olds attending the clinic.

6 (a) $[61.0, 62.5]$

(b) No, both limits greater than 60.

Exercise 5D (page 98)

1 (a) Yes   (b) No   (c) Yes

2 $[0.024, 0.176]$, $[0.238, 0.762]$

3 $[28.5, 39.5]$

4 (a) $[0.144, 0.224]$ (b) Sample is random.

(c) For example, all cars passing under the bridge over a specified period

(d) 49

5 $[0.322, 0.424]$; $n = 971$; variance is an estimate and normal approximation is used.

6 (a) $[0.045, 0.155]$, $[0.159, 0.316]$

(b) Indicates $p_1 < p_2$ since intervals do not intersect, with that for p_1 lower than that for p_2.

Miscellaneous exercise 5 (page 100)

1 (a) $3.357\sigma/\sqrt{n}$ (b) $3.306\sigma/\sqrt{n}$
(c) $3.29\sigma/\sqrt{n}$

These confirm that a symmetric confidence interval has shortest width.

2 $[249.43, 251.39]$

3 (a) 1.791 hours, 0.6615 hours^2

(b) $[1.696, 1.886]$; $[0.149, 0.241]$

No, the confidence interval indicates a mean greater than 10%.

4 (a) 1.04 litres, 0.0192 litres^2

(b) $[1.02, 1.06]$

(c) 520

5 $[0.640, 0.860]$, $n = 180$

6 $[22.8, 24.6]$; normal distribution of sample mean is approximate (central limit theorem) and variance estimate is used.

7 (a) 3.439 (b) 284

8 (a) $[0.0362, 0.0738]$

(b) Between 2710 and 5520 based on confidence interval; 3640 based on p_s

9 (a) Confidence interval for p is $[0.426, 561]$; average.

(b) 95% confidence interval $[0.413, 0.573]$; yes

10 $[1.38, 2.12]$; $P(T < 0) = 0.132$, not small enough; sample large enough for the central limit theorem to apply, so the interval is valid for most reasonable distributions.

6 Hypothesis testing: continuous variables

Exercise 6A (page 104)

1 $H_0: \mu = 102.5$, $H_1: \mu < 102.5$

2 $H_0: \mu = 84$, $H_1: \mu \neq 84$

3 $H_0: \mu = 53$, $H_1: \mu > 53$

4 $H_0: \mu = 30$, $H_1: \mu < 30$. A one-tail test is more appropriate for the customer.

5 $H_0: \mu = 5$, $H_1: \mu < 5$. A one-tail test for a decrease is appropriate since the only cause for concern is μ falling below 5.

Exercise 6B (page 108)

The end-points of the critical regions are given correct to 2 decimal places.

1 (a) Reject H_0 and accept that $\mu > 10$.

(b) Accept $H_0: \mu = 10$.

2 $H_0: \mu = 102.5$, $H_1: \mu < 102.5$; $\overline{X} \leq 101.69$; accept $\mu = 102.5$.

3 $H_0: \mu = 84.0$, $H_1: \mu \neq 84.0$; $\overline{X} \leq 82.11$ or $\overline{X} \geq 85.89$; reject H_0 and accept that mean time differs from 84.0 s.

4 $\bar{x} = 4.87$; $H_0: \mu = 5$, $H_1: \mu < 5$; $\overline{X} \leq 4.71$; accept H_0, mean is at least 5 microfarads.

5 $H_0: \mu = 85.6$, $H_1: \mu < 85.6$; $\overline{X} \leq 78.39$; reject H_0 and accept that the mean is less than 85.6.

6 $H_0: \mu = 6.8$, $H_1: \mu \neq 6.8$; $\overline{X} \leq 6.69$, $\overline{X} \geq 6.91$; reject H_0, and accept that the mean is greater than 6.8.

7 $H_0: \mu = 30$, $H_1: \mu < 30$; $\overline{X} \leq 28.94$; accept H_1, there is cause for complaint.

8 $H_0: \mu = 3.21$, $H_1: \mu \neq 3.21$; $\overline{X} \leq 2.92$, $\overline{X} \geq 3.50$; accept H_0; the sample does not differ significantly from a random sample drawn from the population of all of the hospital births that year.

Exercise 6C (page 111)

1 $H_0: \mu = 330$, $H_1: \mu > 330$; $z = 2.504 > 1.96$; accept H_1, manager's suspicion is correct.

2 $H_0: \mu = 508$, $H_1: \mu \neq 508$; $z = -1.622$. This lies between -1.645 and 1.645, so accept H_0 and the process is under control.

3 $H_0: \mu = 45.1$, $H_1: \mu \neq 45.1$; $z = -2.758$. This is outside -2.054 to 2.054, so accept H_1, the mean has changed.

4 $H_0: \mu = 160$, $H_1: \mu > 160$; $z = 2.546 > 2.326$, so accept H_1, that the mean is greater than 160.

5 $H_0: \mu = 276.4$, $H_1: \mu > 276.4$; $z = 1.328 < 1.645$, so accept H_0, the campaign was not successful. This assumes that the sample can be treated as random; normal distribution of daily sales; the standard deviation remains unchanged.

6 $H_0: \mu = 4.3$; $H_1: \mu < 4.3$; $z = -1.581 < -1.282$, so accept H_1 that the mean waiting time has decreased.

7 $H_0: \mu = 42.3$, $H_1: \mu > 42.3$; $z = 2.594 > 2.326$, so accept H_1, the results are unusually good.

Exercise 6D (page 114)

1 $H_0: \mu = 23$, $H_1: \mu \neq 23$; $z = -2.359$. This is outside the acceptance region -2.326 to 2.326, so reject H_0 and accept $H_1: \mu \neq 23$.

2 $H_0: \mu = 11.90$, $H_1: \mu > 11.90$; $z = 2.5 > 2.326$, so reject H_0 and accept that the mean is greater than 11.90 cm.

3 $H_0: \mu = 5\frac{1}{4}$, $H_1: \mu \neq 5\frac{1}{4}$; $s = 1.582$, $z = 0.554$. This lies between -1.645 and 1.645, so accept H_0, the results are not unusual that April.

4 $H_0: \mu = 375$, $H_1: \mu < 375$; $\overline{x} = 373.85$, $s = 3.800$, $z = -1.914 < -1.645$, so reject H_0 and accept that the mean is less than 375 g.

5 $H_0: \mu = 0.584$, $H_1: \mu > 0.584$; $\overline{x} = 0.6052$, $s = 0.1452$, $z = 1.264 < 1.282$, so accept H_0, melons grown organically are not heavier on average.

6 $\overline{x} = 6.815$, $s = 0.06058$; $z = 1.918$; $H_0: \mu = 6.8$
  (a) $H_1: \mu \neq 6.8$, z lies between -2.24 and 2.24, so accept that $\mu = 6.8$.
  (b) $H_1: \mu > 6.8$, $z < 1.96$, so accept $H_0: \mu = 6.8$.
  (c) $H_1: \mu < 6.8$, $z > -1.96$, so accept $H_0: \mu = 6.8$.

Note that in part (c) where the alternative hypothesis suggests that the population mean is less than 6.8 but the sample mean is greater than 6.8, there is no need to calculate the value of Z. In this situation H_1 cannot be accepted at any significance level.

Exercise 6E (page 117)

All p-values are given correct to 3 decimal places.

1 (a) 0.076   (b) (i) no (ii) yes.

2 $\overline{x} = 8.768$, $s = 2.075$; $H_0: \mu = 8.54$, $H_1: \mu \neq 8.54$; $z = 0.777$, $p = 0.437$. Mean not different from £8.54.

3 $H_0: \mu = 8.42$, $H_1: \mu > 8.42$; $z = 1.703$, $p = 0.044$.
  (a) Mean greater than 8.42,
  (b) mean not greater than 8.42.

4 $s = 0.3084$; $H_0: \mu = 7$, $H_1: \mu < 7$; $z = -1.421$, $p = 0.078$. Accept that mean is less than 7 hours.

5 $H_0: \mu = 31$, $H_1: \mu > 31$; $z = 2.085$, $p = 0.019$ Accept that the mean is greater than 31%.

6 $z = 2.856$; approximately 0.44%. This is the p-value of the test.

Miscellaneous exercise 6 (page 118)

1 (a) $H_0: \mu = 2.855$, $H_1: \mu \neq 2.855$; $Z \leq -1.96$, $Z \geq 1.96$.
  (b) $z = -1.540$, so accept $H_0: \mu = 2.855$, the batch is from the specified population.

2 (a) $\mu < 25$ (b) $z = -2.326$, 1%
  (c) $z < -2.326$, significance level < 1% because $z < -2.326$.

3 (a) $\mu_0 = 210$ (b) 3.9%

4 (a) $z = 2.630 > 1.96$, so accept $\mu > 6.0$, Nisha's average blood glucose level is higher than 6.0.
  (b) $\mu_0 < 6.420$
  (c) (i) valid; sample size large enough for central limit theorem and to replace σ by a value estimated from the sample
  (ii) readings at week-end may be biased by different life style, so results not valid.

5 $\bar{x} = 4.741$, $s^2 = 2.820$. $z = -1.236 > -1.282$, so accept H_0, the mean crop weight per plant is 5 kg. Smallest significance level is 10.8%.

6 $\bar{x} = 98$, $H_0: \mu = 100$, $H_1: \mu < 100$; $z = -2.108 < -1.645$, so accept $\mu < 100$, athlete's performance has improved. The total maximum time is 964.4 s.

7 (a) (i) 10.46 (ii) 15.64
Expected (average) value of estimates of the parameter is equal to the parameter.
  (b) (i) 1.0 approximately
    (ii) Since $10.46 < 10.65$, H_0 is accepted: the mean distance of the houses from the station is not more than 10 miles.
  (c) Part (a) answers only require randomness of sample, so they are still valid. Part (b) has a large enough sample for the central limit theorem to hold, so the results are valid.

7 Hypothesis testing: discrete variables

Exercise 7A (page 124)

1 $H_0: p = \frac{1}{2}$, $H_1: p \neq \frac{1}{2}$, p is proportion of boys; $X \sim B(18, \frac{1}{2})$; $P(X \geq 12) = 0.1189 > 0.05$, so accept H_0, the numbers of boys and girls are equal.

2 $H_0: p = 0.95$, $H_1: p < 0.95$; $X \sim B(25, 0.95)$; $P(X \leq 22) = 0.1271 > 0.05$, so accept H_0, there are at least 95% satisfied customers.

3 $H_0: p = \frac{1}{2}$, $H_1: p > \frac{1}{2}$ (or $\mu = 2.5$ and $\mu > 2.5$); p is the proportion of nails with length greater than 2.5 cm; $X \sim B(16, \frac{1}{2})$; $P(X \geq 13) = 0.0106 < 0.025$, so accept H_1, $p > \frac{1}{2}$ and $\mu > 2.5$.
The symmetry of the normal distribution about its mean is used in the statement $H_0: p = \frac{1}{2}$.

4 $H_0: p = \frac{1}{4}$, $H_1: p > \frac{1}{4}$; $X \sim B(10, \frac{1}{4})$; $P(X \geq 5) = 0.0781 > 0.05$, so accept H_0, the probability that Dinesh predicts the colour of a card correctly is $\frac{1}{4}$.

5 (a) $H_0: p = \frac{1}{6}$, $H_1: p < \frac{1}{6}$; $X \sim B(30, \frac{1}{6})$; $P(X \leq 1) = 0.0295$
  (b) $0.0295 < 0.05$, so the results are confirmed.

6 (a) $H_0: p = 0.7$, $H_1: p \neq 0.7$; $X \sim B(12, 0.8)$; $P(X = 12) = 0.0138 < 0.05$, so reject H_0 and accept that the true figure is not 70%.
  (b) Marie's friends do not comprise a random sample, so test is unreliable.

Exercise 7B (page 126)

1 $H_0: p = 0.3$, $H_1: p \neq 0.3$; $X \sim B(80, 0.3) \approx N(24, 16.8)$; $P(X \leq 19) = \Phi(-1.098)$, since $-1.098 > -1.645$ (or since $0.136 > 0.05$), accept H_0, 30% of the beads are red.

2 $H_0: p = 0.75$, $H_1: p > 0.75$; $X \sim B(150, 0.75) \approx N(112.5, 28.125)$; $P(X \geq 124) = 1 - \Phi(2.074)$; since $2.074 > 1.96$ (or since $0.0194 < 0.025$), reject H_0 and accept that more than 75% get relief.

3 $H_0: p = 0.4$, $H_1: p > 0.4$; $X \sim B(50, 0.4) \approx N(20, 12)$; $P(X \geq 29) = 1 - \Phi(2.454)$; since $2.454 > 2.326$ (or since $0.0074 < 0.01$), reject H_0, the company should adopt the price.

4 $H_0: p = 0.8$, $H_1: p < 0.8$; since $-1.326 > -1.645$ (or since $0.0924 > 0.05$), accept H_0, the deliveries are as stated.

5 $H_0: p = 0.08$, $H_1: p \neq 0.08$; $X \sim B(500, 0.08) \approx N(40, 36.8)$; $P(X \geq 53) = 1 - \Phi(2.061)$; $p = 0.039$
  (a) Reject H_0. (b) Accept H_0.

6 $H_0: p = 0.132$, $H_1: p > 0.132$; $X \sim B(95, 0.132) \approx N(12.54, 10.884\,72)$; $P(X \geq 20) = 1 - \Phi(2.110)$; since $2.110 > 2.054$ (or since $0.0174 < 0.02$), reject H_0 and accept that the drop-out rate is greater for science students.

Exercise 7C (page 129)

1 $X \sim Po(2)$; $P(X \geq 4) = 0.143 > 0.05$, so accept H_0 that $\mu = 2$.

2 $H_0: \lambda = 4$, $H_1: \lambda < 4$, where μ is mean number of accidents per week. $X \sim Po(8)$; $P(X \leq 3) = 0.0424 < 0.05$, so accept that the mean has reduced. $X \sim Po(20)$ including all 5 weeks. Approximate by $N(20, 20)$. $P(X \leq 14) = 1 - \Phi(-1.230)$; since $-1.23 > -1.645$ (or since $0.1093 > 0.05$) the conclusion changes.

3 (a) $n = 60 > 50$ and $np = 4.2 < 5$, so Poisson approximation to binomial distribution applies.
  (b) $H_0: \lambda = 4.2$ or $p = 0.07$, $H_1: \lambda \neq 4.2$ or $p \neq 0.07$; $X \sim Po(4.2)$; $P(X \leq 1) = 0.0780 > 0.05$, so accept H_0, the proportion is 7%.

4 $H_0: \lambda = 1.4$, $H_1: \lambda \neq 1.4$; $X \sim Po(1.4)$; $P(X \geq 4) = 0.0537 > 0.05$, so accept H_0, secretary is probably responsible.

5 $H_0:\lambda = 6$, $H_1:\lambda < 6$; $X \sim \text{Po}(18)$;
   $P(X \leqslant 9) = 0.0154$. At 5% level accept that the
   mean is lower.

6 $H_0:\lambda = 3245$, $H_1:\lambda > 3245$;
   $X \sim \text{Po}(3245) \approx \text{N}(3245,3245)$;
   $P(X \geqslant 3455) = 1 - \Phi(3.678)$.
   Since $3.678 > 1.645$, accept that the mean has
   increased.

Miscellaneous exercise 7 (page 130)

1 (a) $H_0:p = \frac{2}{3}$, $H_1:p > \frac{2}{3}$; $X \sim \text{B}(20,\frac{2}{3})$;
        $P(X \geqslant 17) = 0.0604$, so reject H_0 and
        accept that the proportion is greater than $\frac{2}{3}$.
   (b) Statistical tests are not proofs: they indicate
        the likelihood of a hypothesis being correct.

2 $X \sim \text{B}(25,0.4)$; $P(X \leqslant 5) = 0.0294 > 0.025$ so
   accept that $p = 0.4$.

3 $H_0:p = \frac{1}{3}$, $H_1:p \neq \frac{1}{3}$;
   $X \sim \text{B}(174,\frac{1}{3}) \approx \text{N}(58,38.67)$;
   $P(X \leqslant 51) = \Phi(-1.045)$; since
   $-1.96 < -1.045 < 1.96$ (or since $0.148 > 0.025$),
   accept H_0, the proportion of left-handed
   mathematicians is $\frac{1}{3}$.

4 (a) $H_0:p = \frac{1}{2}$, $H_1:p > \frac{1}{2}$ (p is proportion
        preferring Doggo); $X \sim \text{B}(40,\frac{1}{2}) \approx \text{N}(20,10)$;
        $n = 26$.
   (b) 4.1%

5 $H_0:p = 0.07$, $H_1:p > 0.07$;
   $X \sim \text{B}(125,0.07) \approx \text{N}(8.75,8.1375)$;
   $P(X \geqslant 14) = 1 - \Phi(1.665)$; since $1.665 < 1.881$
   (or since $0.048 > 0.03$), accept H_0 and retain the
   batch.

6 (a) Computers must be lost randomly
        throughout the period at a uniform rate.
   (b) $H_0:\lambda = 18$, $H_1:\lambda < 18$; 1

7 (a) Flaws must occur randomly at a uniform rate
        per metre length.
   (b) $H_0:\lambda = 1.8$, $H_1:\lambda < 1.8$; $X \sim \text{Po}(5.4)$;
        $P(X \leqslant 2) = 0.0948 > 0.05$, so accept H_0,
        the rate has not decreased.
   (c) $X \sim \text{Po}(20.7) \approx \text{N}(20.7,20.7)$;
        $P(X \leqslant 9) = 1 - \Phi(-2.462)$. Since
        $-2.462 < -1.645$ (or since $0.0069 < 0.05$)
        accept H_1, the conclusion changes.

8 (a) $X \sim \text{Po}(1.9)$
   (b) $H_0:\mu = 1.9$, $H_1:\mu \neq 1.9$;
        $P(X = 0) = 0.1496 > 0.05$, so accept H_0,
        there is no decrease.
   (c) $H_0:p = 0.1496$, $H_1:p > 0.1496$;
        $X \sim \text{B}(50,0.1496) \approx \text{N}(7.48,6.361)$;
        $P(X \geqslant 13) = 1 - \Phi(1.990)$; since $1.990 >$
        1.645, reject H_0 and accept that the
        numbers have decreased.

9 $z = -2.069$; either $-2.069 < -2.054$ or $0.0193 <$
   0.02, so reject H_0 and accept that $\mu < 5.00$.

10 $H_0:p = 0.036$, $H_1:p > 0.036$;
    $X \sim \text{B}(500,0.036) \approx \text{N}(18,17.352)$;
    $P(X \geqslant 28) = 1 - \Phi(2.281)$; since $2.281 > 1.645$
    (or since $0.0113 < 0.05$), accept that 0.036 is an
    underestimate.

11 $H_0:\mu = 1.74$, $H_1:\mu > 1.74$; since
    $z = 2.361 > 1.881$ (or since $0.0088 < 0.03$), accept
    $\mu > 1.74$.
    Replace the last 8 words of the sentence by
    '... states that the sample mean is approximately
    normally distributed for large samples even when
    the population is not normal'.

12 $H_0:p = 0.2$, $H_1:p < 0.2$; 0.098. Accept that the
    percentage of defective vases has been reduced,
    since $0.098 < 0.1$.

13 $H_0:\lambda = 6$, $H_1:\lambda > 6$; $X \sim \text{Po}(60) \approx \text{N}(60,60)$.
    $P(X \geqslant 72) = 1 - \Phi(1.485)$. Since $1.485 < 1.645$
    (or since $0.0687 > 0.05$), accept H_0, the
    background radiation has not increased.
    At least 73 particles.

14 (a) 28.9, 83.41
    (b) $z = -1.319 > -1.645$, so accept H_0.
    (c) 106 610.5

15 (a) $H_0:p = 0.8$, $H_1:p > 0.8$; $X \sim \text{B}(20,0.8)$;
        $P(X \geqslant 19) = 0.0692 > 0.05$, so accept H_0,
        the proportion has not changed.
    (b) 6.92%

16 (a) $z = 0.9474$; the rejection region is in the
        negative tail and z is positive.
    (b) 0.172; accept $H_0:\mu = 15\frac{1}{2}$.

17 (a) $\text{Po}(3)$; calls occur randomly, at a uniform
        rate over 2 days.
    (b) $H_0:\lambda = 1.5$, $H_1:\lambda > 1.5$, where μ is the
        mean number of calls per week.
        $P(X \geqslant 5) = 0.1847 > 0.05$ so accept H_0,
        the daily average has not increased.

18 $H_0:\lambda = 0.5$; $H_1:\lambda > 0.5$; $X \sim \text{Po}(0.5)$;
    $P(X \geqslant 3) = 0.0144$; any level $\geqslant 0.0144$.

19 $H_0:p = 0.25$, $H_1:p < 0.25$; $X \sim \text{B}(12,0.25)$;
    $P(X \leqslant 2) = 0.3907 > 0.10$, so accept H_0, the
    student did no worse than by guessing. However,
    the result is less than expected by guesswork.

20 $n > 50$ and $np = 3 < 5$, so the Poisson
    approximation to binomial distribution is
    applicable. $X \sim \text{Po}(3)$; $P(X = 0) = 0.0498 \approx 0.05$.
    $H_0:p = 0.05$, $H_1:p > 0.05$; $X \sim \text{B}(10,0.05)$;
    $P(X \geqslant 2) = 0.0861 < 0.10$, so reject H_0 and
    accept that $p > 0.05$.

21 $H_0 : \mu = 1.005$, $H_1 : \mu > 1.005$; $z = 0.977 < 1.645$,
    so accept $H_0 : \mu = 1.005$.
    $H_0 : p = 0.65$, $H_1 : p < 0.65$;
    $X \sim B(100, 0.65) \approx N(65, 22.75)$;
    $P(X \leq 53) = \Phi(-2.411)$; since $0.0080 < 0.05$,
    reject H_0 and accept that $p < 0.65$.

22   (a) $n > 50$ and $np = 2.4 < 5$, so the Poisson
         approximation to binomial distribution is
         applicable.
      (b) $H_0 : \mu = 2.4$, $H_1 : \mu > 2.4$; $X \sim Po(2.4)$;
         $P(X \geq 6) = 0.0357 < 0.05$, reject H_0 and
         accept that mean has increased.

23 $H_0 : \mu = 3000$, $H_1 : \mu > 3000$; $z = 2.036$; since
    $2.036 > 1.96$ (or since $0.0209 < 0.025$), reject H_0
    and accept that the mean is greater than 3000.
    $H_0 : p = 0.4$, $H_1 : p \neq 0.4$;
    $X \sim B(150, 0.4) \approx N(60, 36)$;
    $P(X \leq 51) = \Phi(-1.417)$; since $-1.417 < -1.96$
    (or since $0.0783 > 0.025$), accept H_0, the
    proportion agreeing with the manufacturer is
    40%.

8   Errors in hypothesis testing

Exercise 8A (page 140)

1   (a) $\bar{X} \geq 6.316$ (b)   0.1963

2   (a) $\bar{X} \leq 9.7419$ (b)   Type I error
      (c)   0.1404

3   (a)   9.78   (b) 0.2039   (c)   It will be smaller.

4   (a) $H_0 : \mu = 1.94$, $H_1 : \mu < 1.94$; rejection region:
         $\bar{X} \leq 1.883$; $\bar{x} = 1.7$, reject H_0 and accept
         that new system had the desired effect.
      (b)   0.0087
      (c)   Accepting that the mean absence rate is 1.94
         when it is actually less than 1.94.

5   (a) $\bar{T} \leq 3.3065$ (b)   0.0877
      (c)   0.0647, which is less than 0.0877

6   (a) $H_0 : \mu = 30$, $H_1 : \mu < 30$; $\bar{X} > 29.7782$
      (b) $0 < \mu < 29.530$

Exercise 8B (page 145)

1   (a)   0.0421   (b)   0.917

2   (a) $H_0 : p = \frac{1}{6}$, $H_1 : p < \frac{1}{6}$
      (b)   10.3%   (c)   0.103   (d)   0.589

3   (a)   0.0635   (b)   0.8593; α is bigger and β is
         smaller.

4   (a) $H_0 : p = 0.1$, $H_1 : p < 0.1$
      (b) $P(X = 0) = 0.0798 < 0.10$, so reject H_0 and
         accept that the modified toaster is more
         reliable.
      (c)   0.0798   (d) $p < 0.0119$

5   (a) $S \geq 95$ (b)   0.018   (c) $p_0 = 0.7875$

6   (a) $X \leq 51$, $X \geq 69$ (b)   0.083
      (c)   Accept that 60% is correct.
      (d)   0.0094

7   (a)   0.0065   (b)   0.2121   (c)   0.0024
    The probability of a Type II error is decreased by
    increasing the sample size.

Exercise 8C (page 149)

1   (a)   0.0430
      (b)   0.9116; accept that the mean is 6.5; Type II
         error if the mean is actually 5.5.

2 $X \sim Po(12)$, $P(X \leq 6) = 0.0458 < 0.05$ and
    accept that mean has reduced; Type I error.

3   (a)   Sample size n is greater than 50 and
         $np = 3 < 5$.
      (b)   One-tail since it would not be concerned if
         the rate were smaller than 5%.
      (d)   (i) Type II   (ii) Type I   (e)   0.703

4   (a) $X \leq 12$, $X \geq 32$ (b)   4.40%
      (c)   The mean has not changed.   (d)   0.928

5   Sales must occur randomly over time and at a
    uniform rate.
      (a) $T \sim Po(16)$; approximate by $N(16, 16)$;
         $T \geq 23$
      (b)   5.21%   (c)   0.856, a large value

6   (a)   0.109   (b)   0.794

Miscellaneous exercise 8 (page 151)

1   (a)   0.1
      (b)   Not possible since Type I error made when
         H_0 is rejected and Type II when H_0 is
         accepted.
      (c)   Type I

2   (a)   The number of misprints $X \leq 1$.
      (b)   Accept that mean has reduced.   (c)   0.0902

3   (a) $B(200, 0.9) \approx N(180, 18)$ since $np = 180 > 5$
         and $nq = 20 > 5$; $k \leq 171$
      (b)   The variable is discrete and only
         approximated by a normal distribution.
         2.3%   (c)   Type II, 0.383

4   (a)   10.1%
      (b)   8.09%

5   (a) $X \sim Po(19)$; approximate by $N(19, 19)$;
         $X \leq 11$
      (b)   0.0426   (c)   0.870

6   (a)   0.4218 mm^2
      (b)   Expectation of the estimate of the parameter
         is equal to the parameter.
      (c) $\bar{X} \geq 7.4755$ (d)   0.297
      (e)   The variance is an estimate.

7 (a) $H_0: p = 0.2$, $H_1: p > 0.2$, where p is the probability of Emma choosing the correct colour.

(b) $X \sim B(20, 0.2)$, $P(X \geqslant 7) = 0.0867 > 0.05$, so accept that Emma is not telepathic.

(c) Type II, when the null hypothesis was true.

8 (a) $H_0: \mu = 75$, $H_1: \mu > 75$

(b) Type I, accept that the regulations are not met when they are; Type II, accept that the regulations are met when they are not.

(c) Type I since ecosystem not affected.

(d) $Z \leqslant 2.425 > 1.282$, so accept H_1, regulations are not met.

9 (a) $H_0: \mu = 12.4$, $H_1: \mu < 12.4$; 0.58%

(b) Approximately 0 (c) $n = 7$, 0.0931

10 (a) $\overline{X} \leqslant 20 - \dfrac{6.41}{\sqrt{n}}$ (c) 0.0429

(d) Decreases (e) 325

Revision exercise
(page 154)

1 0.0161

2 (a) 158.29 grams, 371.23 grams^2

(b) $[367, 375]$ grams
It is not necessary to assume that the population is distributed normally, as the sample is large and so, according to the central limit theorem, the sample mean is normally distributed.

3 (a) $B(50, 0.2)$ (c) $a = 1$

4 (a) 8.01 mm, 0.000 175 mm^2

(b) Test statistic, $z = 2 > 1.96$, so there is significant evidence, just, at the 5% level to accept H_1 that $\mu \neq 8.00$.

5 0.323

6 A measure obtained from the sample where expected value is equal to population variance.

(a) $\mu = 2$, $\sigma^2 = \frac{2}{3}$

(b) Possible samples
$(1,1), (1,2), (1,3), (2,2), (2,3), (3,3)$;
Mean 1, 1.5, 2, 2, 2.5, 3
Distribution of sample mean

Value	1	1.5	2	2.5	3
Probability	$\frac{1}{9}$	$\frac{2}{9}$	$\frac{3}{9}$	$\frac{2}{9}$	$\frac{1}{9}$

E(sample mean) = 2; hence sample mean is an unbiased estimator of population mean in this case.

(c) Distribution of sample variance

Value	0	$\frac{1}{4}$	1
Probability	$\frac{3}{9}$	$\frac{4}{9}$	$\frac{2}{9}$

E(sample variance) = $\frac{1}{3} \neq \sigma^2$.

$E\left(\dfrac{n}{n-1}(\text{sample variance})\right) = \frac{2}{1} \times \frac{1}{3} = \frac{2}{3} = \sigma^2$

so $\dfrac{n}{n-1}(\text{sample variance})$ is an unbiased estimator of σ^2.

7 A random sample of size n is one chosen so that every possible grouping of size n from the population has an equal chance of being chosen. A random sample is chosen to try to avoid getting an unrepresentative sample.
Houses 100 to 150 have no chance of being chosen.
Use three-digit numbers and subtract multiples of 150 until the number is in the range 001 – 150. Ignore 000, 901 – 999 and any repeated selections.
So the sample members would be 88, 52 and 118.

8 (a) 0.0047

(b) It is not necessary for the individual masses to have a normal distribution because the distribution of the mean will be approximately normal (by the central limit theorem) since the sample size, n, is large (100 here).

9 (a) $a = \frac{2}{3}$ (b) $E(X) = 1$ (c) $Var(X) = \frac{11}{45}$

10 $[0.326, 0.410]$

11 (a) $H_0: \mu = 2.00$, $H_0: \mu \neq 2.00$; $z = 3.83$, and the critical region is $|z| > 1.96$. H_0, that there is no change in the setting, can therefore be rejected in this case.

(b) 0.677

12 The events (potholes) occur at random, and they occur independently of one another. The mean number per interval is proportional to the size of the interval. Two potholes cannot occur simultaneously in the same place.

(a) 0.414 (b) 0.00254 (c) 0.205

13 (a) $\frac{3}{8}$ (b) (i) 0 (ii) $\frac{2}{5}$

14 (a) 0.933 (b) 0.986

15 (a) Test statistic, $z = -1.66 < -1.645$, so there is significant evidence, just, at the 5% level that there has been a reduction in service time.

(b) P(Type II error) = 0.786

16 (a) $\overline{X} \leqslant 59.82$

(b) It is accepted that the mean playing time, μ, is 60 minutes, when it is, in fact, less than 60 minutes.

(c) 0.0519

17 Discrete; P_s takes values $\frac{1}{250}r$ for
$r = 1, 2, \ldots, 250$
 (a) [28.1, 39.9]
 (b) Variance is estimate; discrete variable is
   approximated by continuous variable;
   distribution is approximately normal.
 Approximate sample size is 2380;
 99% probability that error is less than $2\frac{1}{2}\%$.

18 $Po(4n)$, this can be approximated by $N(4n, 4n)$
when n is large; $n = 30$.

19 $H_0: p = 0.13$, $H_1: p < 0.13$; $B(2500, 0.13)$ is
approximated by $N(325, 282.75)$, 2.01%; 0.161

20 (a) Easy to do but it may not be simple to find
   the people having selected them. Also, the
   people chosen may not themselves be users
   of soap powder.
 (b) This method ensures that the people selected
   can be interviewed and it can easily be seen
   whether they have bought soap powder.
   Only one shop is being used and it may be
   that people who use smaller or different
   stores have different preferences.
 (c) Some people might be out when the
   interviewer calls. There may also be more
   than one person per household who uses
   soap powder and there may be more than
   one brand used per house.
 (d) This is quick and easy to do but it ignores
   anyone who is not on the phone and so may
   give an unrepresentative sample.

21 (a) $N\left(15, \frac{9}{5}\right)$ (b) 0.932 (c) $7, \frac{1}{5}$
 (d) $N\left(7, \frac{1}{5}\right)$; central limit theorem
 (e) 0.178

22 (a) $H_0: p = \frac{1}{2}$, $H_1: p > \frac{1}{2}$, where p is the
   probability that a car turns right.
 (b) 27
 (c) 0.015

23 (a) 0.0992 grams^2, [16.85, 17.08] grams
 (b) 116
 (c) variance unchanged, confidence interval
   becomes [16.90, 17.13] grams. The answer
   to part (b) is unchanged.

24 0.010, 0.098(5)

Practice examinations

Practice examination 1 (page 160)

1 0.897

2 (i) In a random sample of size n every possible
   set of n items in the population has an equal
   chance of being chosen. Random samples
   are used to avoid bias, so that they can be
   taken as representative of the population
   from which they are drawn.
 (ii) Number the students from 001 to 763; use
   3-digit random numbers (rejecting 000 and
   764 to 999, and ignoring any repetitions) to
   identify 5 students.

3 (i) 3.045, 3.439 (ii) [2.833, 3.257]

4 (ii) $\frac{7}{27}$ (iii) $\frac{3}{4}$

5 (i) 0.0547 (ii) 0.833

6 (i) 0.623 (ii) $Po(33.9)$ (iii) 0.647

7 (i) 0.0861 (ii) 0.0024

Practice examination 2 (page 162)

1 [19.8%, 36.2%]

2 $H_0: p = \frac{1}{6}$, $H_1: p < \frac{1}{6}$, where p is the probability
of a 'six'; since $P(X \leqslant 1) = 0.130 > 0.10$, there
is insufficient evidence to reject H_0, so can't say
that there are too few 'sixes'.

3 $25 \leqslant n \leqslant 48$

4 (i) $Po(3)$ (ii) 0.101 (iii) 25

5 (i) $\frac{1}{4}$ (ii) 0.2601 (iii) 10 800 litres

6 (i) 0.191 (ii) 0.176

7 (i) $H_0: \mu = 2$, $H_1: \mu \neq 2$, where μ is the
   population mean length of the rods
   produced; since $z = 1.88 < 1.96$, there is
   insufficient evidence to reject H_0, so
   accept that the machine is set up correctly.
 (ii) 0.391

Index

The page numbers refer to the first mention of each term, or the box if there is one.

Formulae

Summary statistics

For ungrouped data:

$$\bar{x} = \frac{\sum x}{n}, \qquad \text{standard deviation} = \sqrt{\frac{\sum (x-\bar{x})^2}{n}} = \sqrt{\frac{\sum x^2}{n} - \bar{x}^2}$$

For grouped data:

$$\bar{x} = \frac{\sum xf}{\sum f}, \qquad \text{standard deviation} = \sqrt{\frac{\sum (x-\bar{x})^2 f}{\sum f}} = \sqrt{\frac{\sum fx^2}{\sum f} - \bar{x}^2}$$

Discrete random variables

$$\mathrm{E}(X) = \sum xp, \qquad\qquad \mathrm{Var}(X) = \sum x^2 p - \{\mathrm{E}(X)\}^2$$

For the binomial distribution $\mathrm{B}(n,p)$:

$$p_r = \binom{n}{r} p^r (1-p)^{n-r}, \qquad \mu = np, \qquad \sigma^2 = np(1-p)$$

For the Poisson distribution $\mathrm{Po}(a)$:

$$p_r = \mathrm{e}^{-a} \frac{a^r}{r!}, \qquad\qquad \mu = a, \qquad \sigma^2 = a$$

Continuous random variables

$$\mathrm{E}(X) = \int xf(x)\,dx, \qquad\qquad \mathrm{Var}(X) = \int x^2 f(x)\,dx - \{\mathrm{E}(X)\}^2$$

Sampling and testing

Unbiased estimators:

$$\bar{x} = \frac{\sum x}{n}, \qquad\qquad s^2 = \frac{1}{n-1}\left(\sum x^2 - \frac{(\sum x)^2}{n} \right)$$

Central limit theorem:

$$\bar{X} \sim \mathrm{N}\left(\mu, \frac{\sigma^2}{n} \right)$$

Approximate distribution of sample proportion:

$$\mathrm{N}\left(p, \frac{p(1-p)}{n} \right)$$